Adventures in

# MATHOPOLIS

# ESTIMATING AND MEASURING

**Karen Ferrell,
Catherine Weiskopf,
and Linda Powley
Illustrated by Tom Kerr**

D1166938

BARRON'S

*All inquiries should be addressed to:*
Barron's Educational Series, Inc.
250 Wireless Boulevard
Hauppauge, New York 11788
**www.barronseduc.com**

ISBN-13: 978-0-7641-3867-6
ISBN-10: 0-7641-3867-7

*Library of Congress Catalog No: 2008014256*

**Library of Congress Cataloging-in-Publication Data**

Schultz-Ferrell, Karen.
    Estimating and measuring / Karen Ferrell, Catherine Weiskopf,
Linda Powley.
        p.   cm. — (Adventures in mathopolis)
    ISBN-13: 978-0-7641-3867-6
    ISBN-10: 0-7641-3867-7
        1. Estimation theory—Juvenile literature.   I. Weiskopf, Catherine.
    II. Powley, Linda.   III. Title.

QA276.8.S259      2008
519.5'44—dc22                                          2008014256

Printed and bound in Canada
9 8 7 6 5 4 3 2 1

# Table of Contents

# Chapter 1
# The Mysterious Hooded Stranger

The mysterious hooded stranger fired up his Supersonic Scooter and zoomed away. It wouldn't do for him to get caught now that his evil deed was done.

That would spoil all his plans. After driving a few blocks, he parked his scooter across from his favorite building, the Mathopolis City Hall. The mysterious stranger watched as a police car cruised by on its rounds. It turned a corner and disappeared from view.

Now was his chance.

Whoever popped corn in that machine was in for a big surprise!

Suddenly, the hooded figure heard familiar voices. Inching his scooter behind a huge holly bush, he picked up his gun and crept toward the sound coming from the mayor's office. He peered in through the open first floor window.

Trusty Dusty, the mayor's assistant entered the room with a notepad in hand. "You called, Sir?" he asked.

"I can't find my glasses," Mayor Marbles said.

Trusty pointed to the top of his boss's head.

"Oh," the mayor mumbled as he reached up and pulled them down to the end of his nose. "Now where is that construction order?" He shuffled through the mountain of papers.

"Right here, sir." Trusty pulled a long roll of paper from the safety of his shirt pocket.

"Can't lose that," Mayor Marbles replied. "If we did, our great Mathopolis Fair would be in chaos."

Outside the window, the hooded figure looked down at his Zippo Zappo Eraser gun. "One shot," he thought. "All I need is one clear shot to cause more confusion."

"What's first, Trusty?" the mayor asked.

Trusty read from the list. "Set up twenty-five tables to hold cakes entered in the bake off."

"Are you sure we need twenty-five?" the mayor asked.

"Take a look, sir." Trusty passed the paper to his boss.

"Lumber for the booths and a sign to go across Main Street," read the mayor. He set the list on his desk, and a dreamy expression crossed his face. "Did I tell you about the frog jumping contest and the slug races," he asked, looking like he'd just remembered it was Christmas. "Have you started gathering them up?"

Trusty nodded.

"And we need a theme." The mayor leaned back in his chair and shut his eyes to think.

Trusty frowned. They'd never had a theme before, but if that's what the boss wanted, then a theme it was.

The mayor's eyes flew open. "I've got it," he exclaimed. "We'll have a Bizarre Bazaar."

"Sir?" Trusty said. He wasn't sure he'd heard correctly.

"A Bizarre Bazaar. Anything goes!"

"Like?" Trusty asked.

The mayor thought for a moment, and then stepped out from behind his desk. He was almost dancing with excitement.

"Foot races and outhouse races. Races of all kinds."

Trusty added a note to the bottom of his list.

"And a quilted numbers contest. You know how I love numbers." The mayor went on without even stopping for breath. "We could have a pie-eating contest, only the contestants would have to eat the pies into number shapes."

"My, my," Trusty said, shaking his head. "Is that all?"

"And an ABC bubble gum-sculpting contest."

Trusty groaned, but he kept writing and wondered who would want to work with already-been-chewed gum?

"And we could have sumo wrestlers and camels and all sorts of things."

"If you say so, sir," Trusty agreed. "But camels are hard to come by."

"I have faith in you," the mayor replied. "Let's see what else is on this construction order." He picked up the list from his desk and held it at arms length in front of his face.

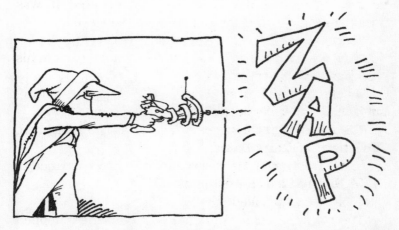

"Hey, there aren't any numbers here," the mayor said, frowning at the list.

"Let me see." Trusty leaned over the mayor's shoulder to look.

"I'll handle this." The mayor grabbed a pen from his desk drawer and filled in all the blank spaces with numbers. "There," he said with a satisfied grin. "That should do it." He handed the list back to Trusty.

"A million nails!" Trusty said.

"Do you think it's enough?" asked the mayor.

The hooded man nodded in satisfaction. Phase One of his plan was off to a great start.

He stole his way to the corner and checked out the street. Other than a scruffy dog sleeping on the steps of City Hall, there was no sign of life.

In a flash, he rescued his Supersonic Scooter from behind the bush and started the engine. He almost holstered his Zippo Zappo Eraser gun, but stopped. The temptation was too great. He pointed the weapon toward City Hall, fired, and sped off down the street.

The dog watched the mysterious stranger disappear from view. He stood, stretched, and then broke into a hurried trot toward the privacy of the park. He stopped when he reached a small, hidden clearing by the pond and looked carefully around. Good. There was no one in sight. Lifting his muzzle to the sky, he let out one long, mournful howl.

7

# Chapter 2
# But It Doesn't Mean a Thing

Elexus Estimator stood next to the scruffy dog. "What's up, Mongrel?"

Mongrel jumped up and placed his paws on either side of a Mathopolis Fair poster stapled to a nearby pole.

"The fair!" Elexus Estimator and Maverick Measurer exclaimed together when they saw the advertisement.

Mongrel barked his approval.

"Funnel cakes," Elexus said, licking her lips. "And salty pretzels and sweet corn and—"

"The fair must be in danger," Maverick interrupted. "And it must be a measuring and estimating crisis or Mongrel wouldn't have called us."

"I'll save the day!" said Elexus.

"You can't estimate everything," Maverick said, hands on his hips. "Measuring," Maverick said, pointing to his tool belt. "The world would collapse without measuring."

"No way," Elexus shot back. "What if something happened to your tools? What if they were lost?" Faster than a Zippo Zappo Eraser gun blast, Elexus unbuckled Maverick's belt and flew off. "Or stolen."

"Hey!" Maverick yelled after her.

Elexus hovered over the park dumpster. "Looks like you'll have to estimate." She waved Maverick's belt over the trash can.

"Don't you dare!" Maverick yelled.

Mongrel flattened his ears and barked.

"You'd better get down here, before he gets really mad," said Maverick.

Elexus swooped down to stand next to the dog. "Sorry, Mongrel," she said, handing the belt back to Maverick. "What's up?"

Mongrel woofed twice, then turned and trotted down the road. Jogging to keep up, Maverick and Elexus followed Mongrel as he led them back to the street in front of Mathopolis City Hall.

"We can fix this in a jiffy," said Elexus. "We'll estimate what all the numbers should be."

"You can't estimate to fix this mess," said Maverick.

"Yes I can."

"No you can't."

"Can, too."

"Can not."

"Can—"

Maverick is right. Measuring is everywhere, from the tick tock of your clock to the scale that your mom broke when it said she'd gained ten pounds. Cooking, running, building, they all involve measuring and plenty of glorious tools.

But Elexus is right, too. Estimating is everywhere. You can use it to check the answers on your math test, to know if you have enough money to buy a soccer ball, or to tell your parents about what time you'll be home from your friend's house.

First things first, though. Whether you are measuring or estimating, you have to work with numbers. There is just no getting around it. To do this, you have to have a good number sense. Never heard of it? Well, your number sense is your sixth sense. If you thought you only had five senses you were wrong.

In short, number sense is knowing how numbers act.

Part of number sense is that you know the numeral one is not just a straight line. It stands for something, like one burp, one cup of magic truth potion that you are forcing your sister to drink after her first date, or one long mile to the nearest bathroom when you are desperate.

Another part of number sense is understanding that a digit and a number are different. The digit is the figure 1.

The digit 1 can make different numbers depending on the place it's in. It can mean the difference between one teaspoon of chili powder and ten teaspoons of chili powder, or between eating one or ten or a hundred jalapeños.

And that brings us to place value, a very important concept for any math. If you don't keep the digits in their correct place, nothing you do will make sense. Place value is extra important for measuring and estimating. Take the number 382.5, for example. Let's say this number stands for the number of slugs that have to be collected for the slug races.

Notice the place where each digit in the number 382.5 is sitting. It's not by chance. But what is the name of each place, and what does it mean? Let's look more closely at the slugs to find out.

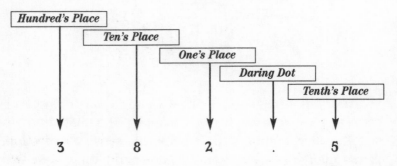

The digit 3 is sitting in the hundred's place. When it sits there, it answers the question, "How many hundreds of slugs are there?" In this case, you have three boxes with one hundred slugs in each, or three hundred slugs; that's what the three in the hundred's place stands for.

The digit 8 is in the ten's place so it answers the question, "How many tens of slugs are there?" You have eight medium boxes with ten slugs in each.

$$8 \times 10 = 80$$

So, you have eighty slugs.

The digit 2 in the one's place answers the question, "How many ones are there?" There are two ones, or two free slugs that are not in boxes.

And then there is the DD, or Daring Dot. The digits after the dot stand for what part of one slug you have left over.

WARNING! DO NOT TRY THIS AT HOME! A PROFESSIONAL STUNT SLUG WAS HIRED FOR THIS ILLUSTRATION. NO ACTUAL SLUGS WERE INJURED.

The digit 5 in the tenth's place tells how many tenths there are, five tenths or one half of one slug.

If you asked the digits to get up and change places, the value of the number would change even though you are using the same digits.

Do you see how the place each digit occupies can make a whole lot of difference with slugs or cups of goop or inches to grow?

**PAUSE FOR THOUGHT**

Okay, pretend you were writing this book and you wanted to have as few slugs as possible. How would you rearrange the digits 8 and 5 and 3 and 2 with only one number to the right of the DD to get the smallest number?

235.8

Here's another place value problem. Fill in the blanks below with the digits 7, 5, and 1.

___ ___ . ___

a. How many different numbers can you make?
b. What is the largest number you can make?
c. What is the smallest number you can make?

a. 6, the possible numbers are 75.1, 71.5, 57.1, 51.7, 15.7 17.5;
b. 75.1; c. 15.7

For all of you who belong to the SPCS (Society for the Prevention of Cruelty to Slugs) here's a less violent example of what each place stands for.

Everyone likes to think about money, so let's consider what $9,563.67 really means.

The 9 means nine 1,000 dollar bills.

The 5 means five 100 dollar bills.

The 6 means six 10 dollar bills

The 3 means three 1 dollar bills.

Of course, there is still the DD, but we don't need stunt money. That's why there are coins.

The 6 means six dimes or 6/10 of a dollar bill.

The 7 means seven pennies or 7/100 of a dollar bill.

Money, slugs, none of this makes sense unless you get the place value concept. So try out your number sense on the following:

1. Given that there are 345 pies entered in the pie-baking contest, which digit tells us how many hundreds of pies there are?
2. If told that there are 56.7 worms left for fishing, which digit tells how many tenths of a worm there are?
3. Given the digits 9 and 5 and 6 and no zeroes, how would you arrange them to get the most money?
4. Who wins the pie-eating contest? Bubba who ate 56 pies or Pearl who ate 65 pies?

1. 3; 2. 7; 3. $965; 4. Pearl

"Now turning to the local news," said the Mathopolis Morning News anchorwoman, "reports of mysterious number vanishings in the downtown area have increased. Several unconfirmed sightings describe a person in a long hooded cloak fleeing the scene. Stay tuned to this station for future updates. And now let's take a look at the weather."

The mayor clicked off the television and sighed. Why did this mysterious troublemaker have to show up now with only two weeks left before the election? Didn't he have enough to worry about?

Trusty stuck his head into the mayor's office. "Sir, there are a couple of young people here to see you."

"Who?" asked the mayor.

Trusty glanced behind himself, and then stepped into the mayor's office and quietly shut the door. "Sir, is it your birthday or anniversary or anything?"

The mayor looked at his desk calendar. "No."

"Then no one would be sending you a singing telegram or something like that today?"

"No." The mayor peered over his glasses at Trusty. "Why do you ask?"

"Well, the people who want to see you are dressed kind of … different."

"Different? How?"

"It's a couple of teenagers in bright costumes," whispered Trusty, as if he thought they could hear him through the door. "And one has a lightning bolt across her chest."

The mayor glanced at his calendar again. "It isn't Halloween." He shrugged. "Send them in. Let's see what they want."

Trusty gave the mayor a doubtful look but opened the door. "The mayor will see you now," he announced.

"Good afternoon, Mayor," said the young lady wearing the lightning bolt. "I am Elexus Estimator, and this is my partner, Maverick Measurer."

"You have popcorn stuck in your hair," said the mayor.

Elexus glared at Maverick. "You said I got it all out."

Maverick grinned mischievously. "I *estimated* that you had."

"Why are you here?" interrupted the mayor.

"We want to help," said Maverick.

"I can always use two good helpers," said the mayor looking them over. "Trusty."

"Yes, sir?"

"Assign them to campaign button duty on the corner of Fifth Avenue and Main Street. Those costumes will attract a lot of attention, and their age will help pull in some of the young adult vote."

"But, sir," interrupted Elexus, "that's not the kind of help we mean. We're here to stop the hooded stranger from eliminating any more numbers."

"How can you help?" asked the mayor.

"We're math superheroes," Maverick replied, puffing out his teenage chest. "Height, weight, volume, and time, if it's not precise, it's a crime."

Not to be outdone, Elexus said, "Volume, time, height, and weight, the answer is to estimate."

The mayor glanced at Trusty.

Trusty nodded his head.

"Welcome to the team," said Mayor Marbles.

Turn the page to see how Maverick and Elexus help Mayor Marbles learn how to measure!

# Chapter 3
## For Good Measure

"What's in the big cardboard box?" asked Mayor Marbles. Before Trusty could stop him, the mayor lifted the lid.

A frog jumped out landing in a mud puddle.

Trusty dropped the box and leapt after it. "They're for the jumping contest," he said, landing in the puddle, too.

The frog jumped away just as one more escaped from the box.

Trusty hopped after them, cornering them near a pile of lumber.

The mayor stifled a laugh. "Who's entering the jumping contest? You or the frogs?"

Wiping mud from his glasses with one hand, Trusty held onto the frogs he'd just captured. "Sir, could you please put the lid back on the box?"

"What? Oh, yes. Of course," said the mayor replacing the lid just before another frog escaped.

The mayor looked around. "For little guys, they sure do move fast." A sudden twinkle lit Mayor Marbles's eyes as he rubbed his hands together with delight. "How about you and I try them out?"

Before Trusty could stop him, the mayor snatched up the box of frogs. "We'll turn one loose, measure how far it jumps, and catch it. Then get another one."

"Who's going to do what, sir?" Trusty asked, although he had a pretty good idea what the answer would be.

"You handle the frogs," the mayor happily replied. "I'll do the measuring."

Trusty nodded and sighed. This was not in his job description. But then neither were a lot of the other things he'd been doing lately.

Meanwhile, lurking behind a partially built dunking booth, the mysterious hooded stranger peered out at them. He fingered his Zippo Zappo Eraser gun in anticipation of another opportunity. It was a small one, but it was still a chance to make the mayor and his faithful assistant feel ridiculous.

He watched as a mud-splattered Trusty captured the other escaped frog. "Ready when you are, sir," said Trusty, placing it on the ground.

The mayor eagerly pulled out his tape measure. "Go!"

Trusty let go of the frog. It sat there staring at the mayor with big bulging eyes.

Trusty waited.

The mayor waited.

The mysterious stranger waited.

"Give it a nudge," the mayor finally said.

Trusty pushed at the frog with the tip of his shoe.

The frog jumped.

The mayor jumped to mark where it landed.

Trusty jumped after the frog.

"Wonderful, marvelous, fantastic," the mayor exclaimed as he stretched his tape measure across the ground.

The mysterious stranger aimed his Zippo Zappo Eraser gun and fired.

The mayor looked down at his tape measure. "That's strange."

"What's strange?" asked Trusty.

"My tape measure," said the mayor. "The numbers are gone. How am I supposed to measure how far each frog jumps without numbers?"

How indeed?

Ever since the first man caught his first frog or bear or fish, there has been a need to measure. Originally, people measured with whatever they could find, like pebbles, sticks, or rocks.

MY FISH IS THREE STICKS LONG.

If you wanted to measure the length of a fish (or a caveman's club) and you didn't have a ruler handy, what would you use?

Whatever tool you choose for measuring would need to be one that didn't change.

Rubber bands wouldn't work very well.

Paperclips, buttons, or jelly beans would work better as long as they were all the same size. (On second thought, jelly beans wouldn't be such a good idea. You might eat them before you finished measuring.)

Long ago people solved the problem of how to measure by using the lengths of their different body parts. No, not nose hairs or eyeballs or belly buttons! They used thumbs, feet, and hands.

For the very smallest objects, they used the width of their thumbs. Try measuring your fish in thumbs. What? You don't have a fish? Measure something else, like a Super Duper Carney corny dog.

What? You don't have a Super Duper Carney corny dog either? Then use a pencil and your imagination.

To measure in thumbs, place your corny dog (also known as a pencil) in front of you. Put your left thumb at the end of your corny dog. Now put your right thumb beside it. That's two thumbs so far.

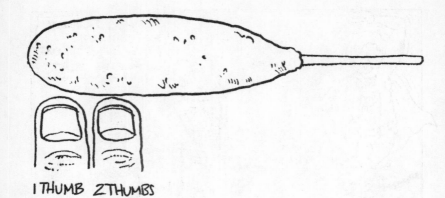

1 THUMB    2 THUMBS

Move your left thumb to the other side of your right thumb. Then move your right thumb to the other side of that. Keep playing Leap Thumb and counting until you get to the end of the corny dog's stick.

Congratulations, you have just used your thumb as a *unit of measurement.*

Thumbs as a unit of measurement would work great for Super Duper Carney corny dogs, but what if you had to measure a longer distance like the distance from the Super Duper Carney corny dog stand to the Hoggily Ever After stand beside it? (Rumor has it they sell really great barbecue.)

You could try to measure with your thumbs, but you'd have to crawl on your hands and knees, people might trip over you, and all the barbecue would be gone before you arrived.

Well, you get the idea. Other people got the idea, too. So for slightly longer things, people used the length of

their foot for measuring. They would place the toe of one foot against the heel of the other, and then repeat the process counting the steps.

Notice that because Maverick's feet are longer than his thumb width (yours are, too, hopefully), he had less counting, less measuring, and less work to do.

So as the distance to be measured increases, so should the length of the unit of measurement you choose to use.

Now suppose Maverick wants to go to the Ice Cream Factory. (There's always room for ice cream.) The factory is way over on the other side of the fairway.

To measure how far away it is, you need another, longer body part as a unit of measurement. For these sorts of distances, people used the length of a belt.

Since Elexus and Maverick aren't the same size, their measurements didn't match!

Answers that didn't match caused a lot of problems a long time ago.

For example, suppose you were in England around the year 1000. You and your fellow knights in shining armor wanted to buy new lances eight feet long. (You were all from the same knight school, and you wanted matching lances for the tournament.) Well, the lance maker charged by the foot.

The lance maker wanted to use *his* foot, which was 9¼ thumbs long.

You wanted to use *your* foot, which was 10½ thumbs long.

You argued.

The lance maker lost the argument, as it were. Mrs. Lance Maker got mad because you killed her husband. She chased you away with a broomstick, so you couldn't get your new lances.

What with lance makers getting killed and angry wives swinging broomsticks, King Henry I decided to do something about this measurement problem. (At least, some historians think it was him.)

Everyone agreed on these measurements:

The width of a thumb was an inch.

The length of a foot was a foot (of course).

The length of a man's belt was a yard.

The problem was whose thumb, foot, and belt should be used?

To solve the problem, King Henry I carved a 12-inch foot on the base of a column of St. Paul's Church in London. He ruled that everyone must use this length as a *foot*. They called this new foot measurement, *by the foot of St. Paul's*.

Each foot equaled twelve thumbs or inches.

The king also made a rule about the length of a yard. He ruled that the yard equaled three feet (the distance from the end of his royal nose to the tip of his index finger when he stretched out his arm).

He even made royal measuring sticks for feet and yards. These *rulers* became the rule for measuring. Today rulers look something like this.

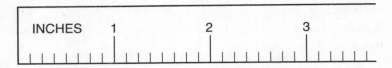

When used, they tell you how many inches or fractions of an inch something measures.

To use a ruler, line the end of the ruler up with the end of what you wish to measure.

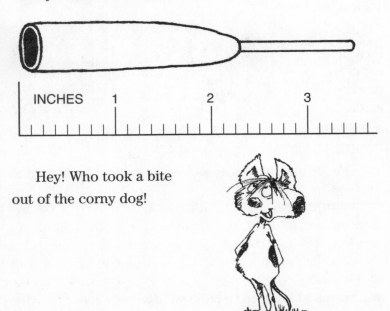

Hey! Who took a bite out of the corny dog!

This corny dog—or what's left of it—and its stick goes past the long three-inch mark. So the corny dog is a little more than three inches long.

Four inches is too much. Three inches is not enough.

To get a more accurate measurement, divide the measurement up into parts. Remember the stunt slug? Remember how it got divided?

To show parts of an inch, marks on the ruler divide each inch into smaller, equal pieces. The inch on this ruler is divided into eight pieces. That means each mark is equal to one-eighth of an inch.

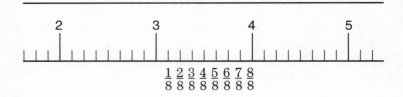

So if the corny dog goes one little mark past the three-inch mark, it is three whole inches plus one part out of eight. You would write it as $3^1/8$". (The " sign is a short way to write inches.) If it goes two marks past, it would be $3^2/8$ inches or $3^1/4$ inches. Four marks would be $3^4/8$ inches or $3^1/2$ inches.

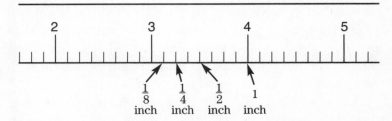

Go back and measure the Super Duper Carney corny dog again. The end of the corny dog is at the half-inch mark between the three and four. So the corny dog is 3½ inches long.

Now use your own ruler and measure the parts of the smiley face to the nearest one-eighth of an inch.

    a. Eyes = _____
    b. Mouth = _____
    c. Head = _____
    d. Hat width (left side to right side) = _____
    e. Width of the hat band (top to bottom) = _____

a. 1/8 inch;   b. 1 inch;   c. 1 1/2 inches;   d. 2 inches;   e. 1/4 inch

Other body parts besides thumbs, waists, and feet have been used for measurements, too. The width of the hand across the fingers is called a *hand* (of course). It's equal to four inches and is still used to measure the height of horses. Similar to that is the *span*. It's the width of the hand with the fingers spread out, about nine inches.

You may have heard the bible story about Noah's ark. The ark was measured in *cubits*. A cubit is the distance from the elbow to the tip of a man's index finger, about eighteen inches.

So:

    1 foot = 12 inches
    3 feet = 1 yard
    36 inches = 1 yard
    1 hand = 4 inches
    1 span = 9 inches
    1 cubit = 18 inches

Now that you know something about measuring short distances, what would you use to measure your fish?

Excellent! You have come a long way since the caveman days.

Mayor Marbles has also solved his problem with a little help from the superheroes.

"Trusty, you mark each step," Maverick instructed. "Mayor, remember to walk heel to toe. Then we'll use your thumb to mark the smaller units."

The mayor vigorously shook Maverick Measurer's hand. "Way to go, Maverick. This ruler is great."

"Thank you, Mayor," said Maverick flashing Elexus a victorious smile.

# Chapter 4
# The Electric Metric

MAYOR MARBLES MIXED-UP
MEASUREMENTS

**INCHES:** 1) SOMETHING YOU SCRATCH
2) ONE-TWELFTH OF A FOOT

**YARD:** 1) NEEDS MOWING ONCE A
WEEK DURING SUMMER
2) THREE FEET OR THIRTY SIX INCHES

**FEET:** 1) WHAT STINKS AFTER
AN HOUR OF GYM CLASS
2) TWELVE INCHES

**HANDS:** 1) WHAT MAYOR MARBLES
HAS TO SHAKE A LOT OF TO GET RE-ELECTED
2) USED FOR MEASURING THE HEIGHT
OF HORSES

**CUBITS:** 1) THOSE BITE-SIZED CHUNKS
OF CHEESE YOU EAT WITH CRACKERS
2) APPROXIMATELY 18 INCHES

Whew! Yards, feet, inches, hands, and cubits! It's hard to keep them all straight.

A lot of other people got confused, too, so the French came up with another system of measuring. It's called the metric system. Once you learn the electric metric trick, remembering metric measurements is faster than lightning.

If you use the measurements in the last chapter, it takes three of this or twelve of that to equal something else. With the metric system, instead of remembering how many inches are in a foot, you only have to remember the number *ten*. All the measurements relate to each other by 10, or 10 × 10 = 100, or 10 × 10 × 10 = 1,000, and more. Knowing this can come in mighty handy sometimes.

First, you need to learn the electric metric trick that makes measuring with metric so easy. As you probably already know, when you multiply a number times ten, the Daring Dot moves one place to the right.

$$5.783.412 \times 10 = 57.834.12$$

Remember Chapter 2? The digits don't change; only their place values change. And when the Daring Dot moves one place to the right, each digit has a value ten times more than before.

Now it's your turn. Multiply these numbers by ten.

a. 8.52 × 10 = _____
b. 0.09 × 10 = _____
c. 189.8 × 10 = _____
d. 5.967 × 10 = _____
e. 147.2213 × 10 = _____

a. 85.2; b. 0.9; c. 1,898; d. 59.67; e. 1,472.213

If multiplying by ten moves the DD one place to the right, how far would the DD move if you multiplied by one hundred?

The answer:

$$5.783.412 \times 100 = 578.341.2$$

See? The DD moved two places to the right, making the number a hundred times bigger than before.

You're getting closer to the lightning-fast trick. Here are a couple of questions to help you figure it out.

How many zeros are there in the number ten?
How many places did the DD move?
Which direction did the DD move?
How many zeros are there in a hundred?
How many places did the DD move?
Which direction did the DD move?

**Do you see the trick?**

YOU MOVE THE DOT
THE SAME NUMBER
OF SPACES AS YOU
HAVE ZEROS.

Now try this one.

5,783.412 × 1,000 = _____

The answer is 5,783,412.

Easy, wasn't it? Now that you know the trick, multiplying by 10, 100, or 1,000 will always be faster than lightning. Use your lightning-fast trick to solve these problems.

    a. $6.82 \times 100 =$ _____

    b. $0.179 \times 1,000 =$ _____

    c. $2.71 \times 10 =$ _____

    d. $0.05624 \times 1,000 =$ _____

a. 682; b. 179; c. 271; d. 56.24

Dividing by 10, 100, or 1,000 is just as easy.

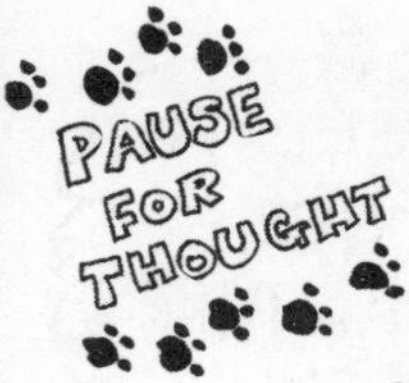

PAUSE FOR THOUGHT

**Can you guess the dividing trick?**

Dividing by ten also moves the Daring Dot one place, but since dividing in this case makes the number smaller, this time the DD moves to the left.

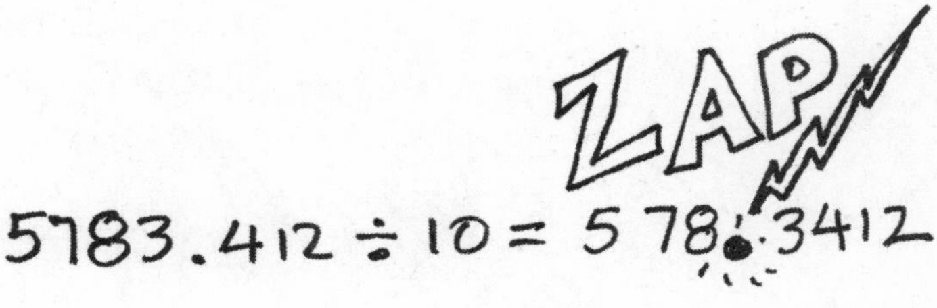

ZAP
5783.412 ÷ 10 = 578.3412

Now each digit has a place value ten times less than it had before. Try using the lightning trick for the following:

a. 53.47 ÷ 100 = __________
b. 8,913.2 ÷ 1,000 = __________
c. 45.32 ÷ 10 = __________

a. 0.5347; b. 8.9132; c. 4.532

So what happens if you have to move a decimal so far left or right that you run out of digits? Simple. Just zap in some zeros.

If you are moving the DD to the right (multiplying), you'll zap the extra zeros onto the right, at the end of the number.

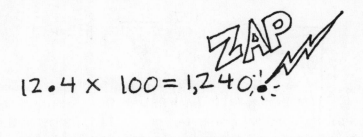

$12.4 \times 100 = 1,240.$

If you're going to the left (dividing), zap zeros onto the left, at the beginning of the number.

$12.4 \div 1,000 = .0124$

Now try the following:

a. $67 \times 1,000 =$ _____  d. $8 \div 100 =$ _____
b. $234 \times 100 =$ _____  e. $0.9 \div 10 =$ _____
c. $9.1 \times 1,000 =$ _____  f. $7.81 \div 1,000 =$ _____

a. 67,000  b. 23,400  c. 9,100  d. .08  e. .009  f. .00781

So what does all this zapping have to do with measuring and the metric system? Instead of three of this or twelve of that, all the measurements relate to each other in groups of ten just like your fingers, toes, and numbers.

So let's group some metric measurements.

The smallest metric measurement is the millimeter. It is very, very thin. What is the thinnest thing you can think of?

Actually a millimeter (mm) is about the width of a needle. Millimeter marks can be found on the opposite side of most twelve-inch rulers.

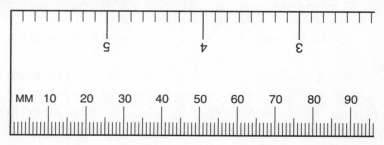

The tiniest marks on a metric ruler are the millimeters.

Now measure your Super Duper Carney corny dog from Chapter 3 again. This time use millimeters.

Just count all the tiny little marks from the beginning of the corny dog to the end of the stick.

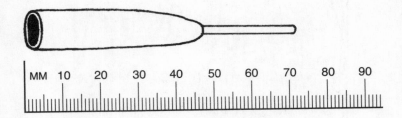

This Super Duper Carney corny dog is 72 millimeters long, but counting every single little mark was boring. So let's group some millimeters and add some lightning speed to the measuring.

Ten millimeters grouped together equals one centimeter (cm). On a ruler, centimeters are the longer marks with the numbers next to them.

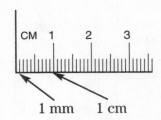

1 mm    1 cm

What can you think of that might be a centimeter wide?

A centimeter (cm) is actually just a little smaller than the width of a dime.

Measure the Super Duper Carney corny dog again, using centimeters this time.

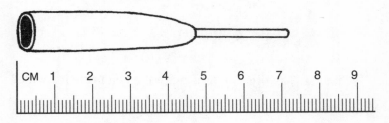

There are seven whole centimeters plus two little millimeter marks. So that makes 7.2 centimeters. The DD marks where the centimeters ended and the millimeters began. You could also have changed from millimeters to centimeters by dividing by ten.

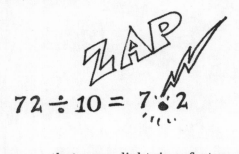

$$72 \div 10 = 7.2$$

Either way, that was lightning fast compared to counting all the little tiny millimeter marks.

The next biggest measurement is the decimeter (dm). It is (you guessed it) a group of ten centimeters.

INTRODUCING DECIMETER!

- LARGER THAN A CD CASE, SHORTER THAN A DOLLAR BILL, AND TEN TIMES LARGER THAN CENTIMETER!

A decimeter is actually about the width of your hand across the fingers. The decimeter isn't labeled on the ruler, but can you find it?

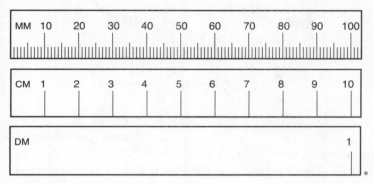

*Not actual size.

Now, take another look at the Super Duper Carney corny dog. This time think decimeters.

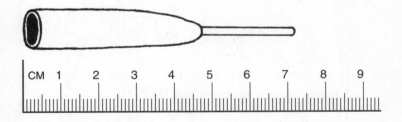

The ten-centimeter mark makes one decimeter. But the corny dog doesn't make it to one decimeter. We can only describe it as .72 decimeter.

Grouping ten decimeters together makes one meter (m).

INTRODUCING

MR. METER!

- LONGER THAN A YARD,
- SHORTER THAN A JUMP ROPE, AND TEN TIMES **LARGER** THAN DECIMETER.

A meter is much larger than this book, so it won't fit inside without folding it up. But you don't need to measure. You can use what you know so far to figure it out. Take a look at these measurements again:

72 mm
7.2 cm
.72 dm

How far did the DD move each time the unit of measurement got bigger? Correct! One space. So, how far will the DD need to move to change from decimeters to meters?

.72 dm = .072 m

Correct again. Another space. Zap! Quick as lightning the DD moved one more place to the left and changed from decimeters to meters.

Okay, take a deep breath.

Now try going from a larger unit of measurement to a smaller one. Take another deep breath and think about it. If the Daring Dot moved left as the units got larger, it will move right as the unit of measurement gets smaller.

Here's an example.

| | |
|---|---|
| The height of the average eleven-year-old | 1.465 m |
| Move the DD one place to the right (1.465 × 10). | 14.65 dm |
| Move the DD one place to the right again (14.65 × 10). | 146.5 cm |
| Move it one more time (146.5 × 10). | 1,465 mm |

That's all there is to it.

So, now that you know something about measuring with metrics, what would be the best metric unit of measure to use for these? Use the millimeter, centimeter, and decimeter rulers to help you.

1. The height of a doorway
2. The width of a donut
3. The size of a donut hole
4. The thickness of a coat hanger

5. The size of your pillow
6. A picture frame
7. The thickness of the piece of spaghetti
8. The length of a piece of spaghetti

**1. meters; 2. centimeters; 3. centimeters; 4. millimeters; 5. meters or decimeters; 6. decimeters or meters; 7. millimeters; 8. centimeters or decimeters**

Here's one more lightning-fast trick to remember. Did you notice that all these metric measurers have the same last name? They all end with the word *meter*. Meter in this context means all these units are used to measure length. The milli, centi, and deci are added to the front of *meter* to show its size.

Milli means "thousandths."
Centi means "hundredths."
Deci means "tenths."

The metric system uses these prefixes again for other kinds of measures. Since you have learned about this metric family, you'll be able to use these prefixes over and over as you explore other types of measuring in the next few chapters.

# Chapter 5
# Wanna Race

The hooded stranger peered out from his hiding place under the grandstand and scowled. Foiled again. So far that meddlesome Maverick Measurer had interrupted all of the stranger's best ideas for Zippo Zappoing things. Still, he had managed to be partly successful, and he was only in Phase One of his plan. He wasn't going to give up—not yet. He pulled his cloak tighter around himself and crept closer to where Crusty or Musty or What's-his-name was talking with Elexus and Maverick.

"Watch out," Trusty warned. "Camels spit."

Elexus backed away from her camel a little.

"Now that we have camels," said Trusty consulting his list, "we need to measure out a race course for them."

"I can do it," volunteered Elexus as she tied her camel to a stake.

Maverick shook his head and puffed out his chest. "No way. You can't estimate a race course," he said securing his camel next to hers. "This is a job for me. We have to be exact. And it isn't a job for inches, feet, or yards. It's a job for bigger measurements."

In the olden days, way before your parents were born, the ancient Romans needed something to measure longer distances, too, like the distances between cities. Since they had a lot of military people and marched everywhere, they decided on what they called the *pace*. A Roman pace equaled two steps of a marching soldier, and it was a little less than five feet.

Do you think your pace is that long? Let's see.

Suppose you got dressed in the dark and accidentally put on one striped sock and one polka-dotted sock. As you walked, the distance from one striped-sock step to the next striped-sock step (say "striped-sock step" three times fast) would be your pace.

Today we keep track of something similar to the Roman pace when we use a pedometer. Instead of paces, it uses a measurement called a *stride*. A stride is one step instead of two, and is the distance between your back foot's toe and your front foot's toe.

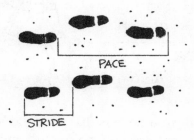

PACE

STRIDE

To get a good measure of your average stride, walk ten steps, measure the distance from the first step's toe to the tenth step toe. Then divide this distance by ten to get your stride.

Each time you take a step wearing a pedometer, it counts your stride. To figure out how far you walked, multiply your stride length by the number recorded on the pedometer.

Here's a little practice. Pick who walked farther, a or b.

1. a. Joe has a stride of 18 inches and walked 50 steps.
   b. His little sister, Amy, has a stride of 15 inches and walked 65 steps.
2. a. Lavita has a stride of 17 inches and walked 125 steps.
   b. Lavita's father has a stride of 24 inches and walked 110 steps.
3. a. Jamal has a stride of 19 inches and walked 137 steps.
   b. Boswick has a stride of 18 inches and walked 142 steps.

1. b; 2. b; 3. a

Say you wanted to measure the distance from your house to school. A pace or a stride wouldn't be much better than a yard stick, although it would be a great excuse for coming to school in mismatched socks.

The Romans had the same problem (not the mismatched socks thing, the longer measurement problem).

**What unit of measure do you think the Romans used for measuring these longer distances?**

If you said the *mile*, then you are correct. A Roman mile was equal to 1,000 paces. (Now that's a lot of marching!) So, if each Roman pace is about five feet and there are 1,000 paces in a Roman mile, how many feet were in their mile?

As with other measurements, though, the length of the pace and the length of a mile have changed since ancient Roman times. Today's mile is 5,280 feet long. In Chapter 3, you learned that there are three feet in each yard. So, how many yards are in a mile?

Fortunately, you don't have to use a yardstick to measure miles. Devices that will measure miles for you are called odometers. Odometers have actually been around a long time. In fact, Benjamin Franklin had one attached to his carriage.

In today's cars, the odometer is usually located in the dashboard, just in front of the steering wheel. It is a row of numbers that changes as the car moves. Some bicycles have odometers, too, as do other vehicles with wheels or tires.

With cars, tires are the key. The odometer works by using the distance around a tire to figure out how many miles a car has traveled. The distance around a circle (or a tire) is called the circumference. One revolution of the tire (one circumference) equals the distance the car has traveled after the tire has gone around once.

The circumference of the tire is programmed into an odometer (if it is computerized) or figured into the mechanics of the unit (if it is mechanical). Just as a pedometer counts the number of steps, the odometer counts the number of times a tire goes around. Instead of you having to do the math, however, the odometer automatically multiplies the number counted by the circumference. This number is then changed into miles.

For example, suppose a tire has a circumference of seven feet. Every time the tire goes around once, the vehicle travels seven feet.

If this tire goes around 100 times, how far will it travel?

**7 feet × 100 = 700 feet**

How many miles will that be?

**700 feet ÷ 5,280 = 0.1326 miles**

So how many times would a tire with a seven-foot circumference have to go around to cover a distance equal to one whole mile?

5,280 FEET IN A MILE ÷ 7 FEET PER REVOLUTION = ABOUT 754 REVOLUTIONS

That means a tire would have to go around 754 times in order to travel just one mile. In other words, they work very hard to get people where they want to go.

Now that you have added miles to your measuring tools, look at the list below. Would you use inches, feet, yards, or miles to measure these?

1. A pig
2. A quilt
3. A cherry pie
4. A Ferris wheel
5. The course for a bike race
6. How far you have to throw a dart to hit the bull's-eye
7. The dart you just threw
8. The height of a circus tent

1. feet; 2. feet; 3. inches; 4. yards??
5. miles; 6. yards; 7. inches; 8. yards

So now you know that

1 mile = 5,280 feet
1 mile = 1,760 yards

The number 5,280 is a really big number to remember in order to change from feet to miles. To make long distances a little easier, metrics comes to the rescue again.

In the last chapter, you met the smaller members of the meter family, Milli Meter, Centi Meter, and Deci Meter. Now meet their big brother.

INTRODUCING
**KILOMETER!**

- **LONGER** THAN TEN FOOTBALL FIELDS!
- **FARTHER** THAN TEN HOME RUN BALLS!
- **MORE THAN TEN LAPS AROUND** YOUR SCHOOL'S BASKETBALL COURT!

Both miles and kilometers measure long distances, like how far you've gone on a road trip. The numbers for kilometers can usually be found under the numbers for miles on a speedometer. A speedometer tells how fast a car is going in *miles per hour* (mph) or *kilometers per hour* (kph) depending on which measurement is used.

It has a pointer in the middle that rotates to point at larger numbers as the car goes faster. So, suppose the speedometer arrow is pointing to the forty-five. That means if the car keeps going at that speed for one hour it will have traveled forty-five miles, thus the term *miles per hour*.

Now look closer at a speedometer that has both miles and kilometers on it. Notice that the kilometer numbers get bigger faster than the mile numbers. That's because kilometers are shorter than miles. It takes more than one and a half kilometers to equal one mile.

How far away would you like to live from the following:

1.  Your best friend
    a. 1 km         b. 1 mi

2.  Your worst enemy
    a. 2 km         b. 2 mi

3.  A skunk farm
    a. 3 km         b. 3 mi

4.  The world's best free amusement park
    a. 4 km         b. 4 mi

1. a; 2. b; 3. b; 4. a

But you can't drive a car to the moon or across the ocean to measure how far away things are. There are other ways to measure distance in space or across the water.

A *nautical mile*, or a sea mile, measures the distance across water. In the English system it is 1.1508 miles or 6,076 feet.

Astronomers measure distances in space in *light years*. A light year is exactly what it sounds like: how far light can travel in one year. Astronomers use this measurement when they are looking at the stars. A light year is about 5,865,696,000,000 miles. That's a lot of steps!

Since camels prefer the desert, Maverick won't need to measure in nautical miles for the camel race. Today's camels can't fly to the moon either.

"I'll measure the distance around the track for the camel race in land miles," said Maverick. "But I'll need a car."

"You can't drive," said Elexus.

Maverick looked around the fair grounds. "I can drive this," he said jumping into a nearby golf cart. "The keys are here."

"Be careful," warned Trusty. "We need that cart to get around the fair grounds."

"Nothing to it," said Maverick, turning the key.

As the engine sputtered to life, the mysterious hooded stranger crept closer and crouched behind the camels.

Maverick puttered up to the starting line. He glanced at the cart's odometer. "The cart's gage is at 348.5 miles. I'll whip around the track once to see how far it is. Then, with a little math, we can figure out how many times the camels have to go around the track for a total of two miles," he explained as he put the car into gear. "Want a ride, Elexus?"

"No thanks," she said, crossing her arms. "I'd rather fly."

"Suit yourself." Maverick pushed the accelerator. He wished for a cloud of dust but none appeared.

He circled the track. Arriving back at the starting line, he jerked to a halt. "Now, we'll take a second look at the odometer."

The hooded stranger smiled to himself, aimed his gun, and fired.

Maverick glanced at the odometer. "The track is exactly—" He stopped in mid-sentence. "Hey! This can't be right. This thing has all zeros!"

The hooded stranger grinned in triumph.

"But no problem," said Maverick. "It makes everything easier. I won't have to do any subtracting now to measure the distance." With that he pushed down on the accelerator and puttered away.

The hooded stranger jammed his Zippo Zappo Eraser gun back into its holster. Drat that meddlesome Maverick. Phase One of his plan hadn't created as much chaos as he'd hoped, but time was running out. Ready or not, he was going to have to move on to Phase Two. He slunk back toward the shadows, but not before a stream of camel spit hit the back of his hood.

Will the hooded stranger continue to cause chaos? Read on and see!

# Chapter 6
# Weight Up

"Okay," said Mayor Lostis Marbles clapping his hands. "Let's get organized."

Sumo-wrestler contestants packed the huge meeting room from corner to corner. Trusty shuddered and sank a little lower in his seat beside the podium. He was glad he didn't have to tell these huge guys what to do. Skulking behind a particularly large wrestler, the hooded stranger listened with interest.

"I think the easiest way to match everyone up," continued the mayor, "would be alphabetically. Are Alexander Aardville and Buzzy Brandish here?"

The crowd parted as an enormous man lumbered to the front of the room. "I'm Alexander Aardville," rumbled the huge man, "but everyone calls me The Axe."

A second, much smaller, man pushed his way to the front. "I'm Buzzy Brandish," he said, nervously eyeing the huge man beside him.

"And what does everyone call you?" asked the mayor kindly.

"The Bee," he replied, still keeping a wary eye on the giant beside him.

"The Axe and The Bee. Excellent," mumbled the mayor making a note on his long list. "You two will be matched in the first round."

The mysterious hooded stranger holstered his gun and slunk away. He didn't need to zap anything this time. The mayor was doing a fine job making a mess of this all by himself.

Mayor Marbles may not think that weight matters in wrestling, but The Bee certainly does, and so did your parents when you were born. People asked your parents lots of questions. One question probably would be, "How much does your baby weigh?" You were teeny tiny back then, so the doctors and nurses weighed you with teeny tiny measurements.

In the weighing system that is used in the United States, one of the smallest measurements for weighing things is ounces. Of course, you're not a thing, and you weren't small enough to be weighed in ounces, but some things that are small enough to be weighed in ounces are:

Candy bars
A package of pencils
A deck of playing cards
A bag of marbles
A baseball

Notice that these things are all small and don't weigh
very much.

Not everyone wants small. Sometimes you want larger
amounts.

DOGGIE GOODIES

To weigh slightly larger amounts, the pound is used. Sixteen ounces equals one pound. Many things weighed in pounds are sold in grocery stores.

Bigger packages of candy
Fruit
Meat
Deli food (potato salad, cold
    cuts, and so on)
Cereal

You can find scales for weighing fruits and vegetables in a grocery store. The scale helps customers weigh what they are about to buy. More often, stores sell prepackaged food with a label on the outside telling the weight and the price.

Hardware stores also sell nails and screws by their weight. For small jobs, there are small packages of nails, screws, bolts, and washers weighed in ounces. For bigger jobs, there are large buckets or boxes of nails weighing five, ten, and twenty-five pounds or more.

All stores want to get paid for every ounce of an item, so they use a combination of pounds and ounces to get the most exact weight. For example, a package of meat might weigh three pounds, seven ounces. It would be written like this on the package, 3 lb. 7 oz.

Notice that *pounds* is abbreviated as *lb*.

Ounce is also abbreviated strangely as *oz*.

Now that you know something about pounds and ounces, go on a field trip to the grocery store and try this grocery store scavenger hunt.

In the produce department, look for each of the items listed. Are they sold by the pound? Are they already bagged and labeled or do you get to bag your own? Which ones are priced by the number you buy instead of by the number of pounds? Put checkmarks in the chart to help you remember what you discovered. If they aren't prepackaged, put a few items in a bag and practice weighing them for yourself. (Hint: Sometimes an item is sold in more than one way.) Add a couple of your favorite fruits or vegetables to the chart.

| Item | How It Is Sold | | | |
|---|---|---|---|---|
| | Bag Your Own | Already Bagged | Pounds | Ounces |
| Grapes | | | | |
| Apples | | | | |
| Potatoes | | | | |
| Onions | | | | |
| Lettuce | | | | |
| Celery | | | | |
| Cantaloupe | | | | |
| Asparagus | | | | |

NO, MONGREL. STEAK
ISN'T A FRUIT OR VEGETABLE.

Look for weight labels in other parts of the grocery store. What can you buy that weighs exactly one pound? (Remember that 16 oz also equals 1 lb.)

YES, MONGREL. STEAK
IS SOLD BY WEIGHT.

Be warned!

There is also a measurement called fluid ounces used for measuring the volume of liquids. Things like drinks, canned soup, and liquid cooking oil use this ounce measurement instead of the ounce for weight measurement. You will learn more about fluid ounces in the next chapter about volume.

Let's get back to talking about you.

When you were born, your parents measured exactly how much you weighed. Baby weight is very important. There are special scales just for weighing babies. Some even have a baby seat attached so the baby is more comfortable.

Of course, not all babies weigh the same. Some babies get really big.

As you got bigger, you still got weighed, but the ounces were not as important anymore. You were measured in pounds. You probably have a scale for weighing people in your bathroom.

For some people, their weight is very important to their job. For example, jockeys get weighed every time they ride a horse in a race.

Weight lifters and wrestlers get weighed before every event. They are matched for each competition according to their weight in order to make the contest fair.

Just as you got bigger, baby animals grow up too. Veterinarians have special scales to weigh everything from kittens to horses. Farmers have scales for weighing cattle and other farm animals. It has a cage on it to keep the animals from jumping off the scale. Some of these scales hold and weigh up to 10,000 pounds of animals. That's a lot of pounds and a lot of zeros.

For weighing huge amounts (like 10,000 pounds of farm animals), the ton is used. One ton is equal to 2,000 pounds. So if the farmers' scale can weigh 10,000 pounds of cattle, how many tons does that equal?

$$10,000 \div 2,000 = 5 \text{ tons}$$

Depending on the breed, a steer can weigh more than a ton. But the largest land animal in the world is the elephant. That not-so-little baby elephant can grow up to weigh 14,000 pounds, about the same as a school bus. What would that be in tons?

14,000 ÷ 2,000 = 7 tons

## A Zoo Quiz

An adult male (bull) elephant eats 170 lb. of hay and feed a day.

1. How many days will it take for a bull elephant to eat a ton of food?
2. How many pounds of food would a bull elephant eat in a year? (Hint: There are 365 days in a year.)
3. How many tons of food is that?

3. 62,050 ÷ 2,000 = Over 31 tons
2. 170 × 365 = 62,050 lb
1. 2,000 ÷ 170 = 11 3/4 days

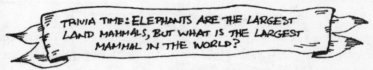

TRIVIA TIME: ELEPHANTS ARE THE LARGEST LAND MAMMALS, BUT WHAT IS THE LARGEST MAMMAL IN THE WORLD?

The great blue whale weighs up to 1,500 tons.

To summarize the weight measurements you've learned so far:

> 16 ounces (oz) = 1 pound (lb)
> 2,000 pounds (lb) = 1 ton

A king-sized riddle: What did the elephant take on vacation?

To find the answer, first fill in the blanks for each of these questions.

H. 4 lb = _____ oz

I. 32 oz = _____ lb

K. 10,000 lb = _____ tons

N. 6 tons = _____ lb

R. 1 ton = _____ oz

S. 10 lb = _____ oz

T. 88 oz = _____ lb

U. 9,000 lb = _____ tons

Now put the letters that match your answers in the boxes below.

| 64 | 2 | 160 | 5½ | 32,000 | 4½ | 12,000 | 5 |
|----|---|-----|-----|--------|-----|--------|---|
|    |   |     |     |        |     |        |   |

His trunk.

Just as distance has a metric measurement called meters, weight has a metric system measurement.

The gram is a very small measurement. To give you an idea of how small, one piece of M & M candy is equal to about one gram.

Here are the weights of some other things in grams.

The smallest weight measurement in the metric system is the gram's little sister, Milli. One *milligram* is equal to 1/1,000 of a gram. To put it another way, 1,000 milligrams are equal to one gram. So to change from grams to milligrams, multiply times 1,000.

For example, the bite-sized Snickers would be equal to

9 grams × 1,000 = 9,000 milligrams

The piece of candy corn would be equal to

1.7 grams × 1,000 = 1,700 milligrams

So how many milligrams would each of the following weigh:

    a. The sheet of notebook paper
    b. The sour ball
    c. The wooden pencil
    d. The Reese's Peanut Butter Cup

a. 3,000, b. 3,000, c. 5,200, d. 8,000

To see just how small a milligram is, take one M & M and cut it up into 1,000 equal pieces.

Too hard and messy?

Then try this. The sheet of notebook paper weighs 3 grams or 3,000 milligrams. Cut the paper up into 3,000 equal pieces. One of those pieces would weigh one milligram.

As you may have guessed, it isn't very practical to try to weigh something in milligrams. Most of the time grams work much better.

The big brother of grams is Kilo Gram. The *kilogram* is equal to 1,000 grams. So 1,000 M & M's would weigh one kilogram.

Who uses the metric system?

In the United States, scientists measure chemicals using the metric system. The tiny milligram allows them to measure liquids and powders for experiments more accurately than they could with ounces or pounds.

Since accuracy when mixing medicines is very important, pharmacies also use milligrams to measure the ingredients for different drugs.

Now that you have learned about two of the metric system families, you may have noticed the metric system trick that makes it easy to remember. All the members of the measuring families have the same first names. The last name changes to show what they measure.

|  | Length: The Meter Family | Weight: The Gram Family |
|---|---|---|
| Milli (1/1,000) | Millimeter | Milligram |
| Kilo (1,000) | Kilometer | Kilogram |

So how do milligrams, grams, and kilograms compare to ounces, pounds, and tons?

To answer this very important question, go back to the store. Many products, whether bagged, boxed, or packaged, have both kinds of weight measurements on their container. Here are some examples.

| Item | Ounces/ Pounds | Grams |
|------|----------------|-------|
| A package of artificial sweetener | 0.035 oz | 1 gram |
| A 6-pack of cheese butter crackers | 1.25 oz | 35 grams |
| A bag of Peanut M & M's | 1.74 oz | 49.3 grams |
| A box of gelatin dessert | 3 oz | 85 grams |
| A jar of crunchy peanut butter | 1 lb 2 oz (18 oz) | 510 grams |

Notice that it takes a lot of grams to make one ounce. To be exact, one ounce equals 28.35 grams. Since ounces are larger than grams, to change the unit of weight from ounces to grams, multiply the number of ounces by 28.35.

Try testing these measurements to see how accurately the packagers converted from ounces to grams. Here's an example to get you started.

1 package of artificial sweetener = 0.035 ounces
0.035 × 28.35 = 0.99

Ah ha! The answer is actually a little less than one gram. Try converting the other items in the chart to grams and see how accurate they are.

crackers = 35.44 grams; M&M's = 49.33 grams; gelatin = 85.05 grams; peanut butter = 510.30 grams

What if you want to go the other way and change the grams to ounces? To change from a smaller unit of measure to a larger one, you would divide. So to reverse what you did above, divide the number of grams by 28.35.

Try converting these gram weights to ounces.

a. 500 grams
b. 324 grams
c. 12 grams
d. 82 grams

You could also change pounds to grams using multiplication. One pound is equal to 453.60 grams. So, to convert from pounds to grams, multiply the pounds by 453.60 grams. To convert from grams to pounds, multiply by 0.002 pounds per gram.

Changing between metrics and pounds or ounces isn't exactly easy! You may have noticed that the answers to these exercises stopped after the hundredths place. That's because the answers didn't come out even. They just kept right on going on forever. The answers were rounded. You'll learn more about rounding answers later in this book.

It's time to put these measurements to use! Turn the page to see Maverick help Mayor Marbles bake a cake!

# Chapter 7
## Too Many Cooks

"How is our City Hall cake coming for the contest?" the mayor asked Trusty.

"Sir?" This was the first time Trusty had heard about a cake. He reached down and absently scratched Mongrel on the head.

"We must have a cake," Mayor Marbles declared. "All the city departments are entering the contest. The policemen have one shaped like a badge. The firemen made a huge red fire truck cake that squirts real whipped cream. Why, even the dog pound made one, using crumbled dog food."

Mongrel licked his lips.

"I have my grandmother's recipe," Mayor Marbles said, pulling a worn looking index card from his hip pocket. "Can you take it to City Hall's Baker Bob?"

"It's Baker Bob's day off," Trusty reminded the mayor.

"Then we'll just have to do it ourselves." The mayor looked at the index card.

Recipe

Ingredients:
2 large eggs
2 c. sugar
2 c. flour
3 tbsp. cocoa
1 c. margarine
1 tsp. vanilla
1 tsp. soda
½ c. buttermilk

The mayor frowned. "I can't make any sense out of this. Is it written in code? What is a *tsp*?"

Trusty started to explain.

"And what's the difference between a large egg and a small one?" The mayor scratched his head. "This is important. We can't afford to make a mistake. I must form a committee to examine this issue, and the sooner the better, so we can get the cake baked."

Trusty had his doubts about a committee, but if that was what the mayor wanted, then—

"Mayor! Trusty! Over here," called the volunteers setting up the pie-eating contest. "Quick!"

The mayor motioned for Trusty to follow, and the two took off running, neither noticing Mongrel as he slipped the recipe from the mayor's hand.

Maverick took the recipe card from under Mongrel's foot and read the instructions.

"No problem," he said when he was through. "I can do this."

"You?" cried Elexus, "What do you know about cooking?"

"More than you do." Maverick held the card in front of her face. "I know exactly which measurements to use."

"Who needs those?" Elexus replied, waving Maverick away. "All I need to know is what goes into the cake. Exact amounts don't matter."

"Wanna bet?" Maverick asked. "Let's see who can make the best cake."

"You're on," Elexus said.

What Trusty tried to explain to the mayor was that the recipe wasn't really written in code. It was written with abbreviations, a short form of a real word.

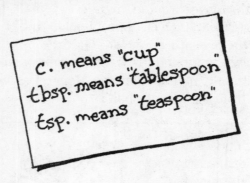

c. means "cup"
tbsp. means "tablespoon"
tsp. means "teaspoon"

And all these terms are used for measuring the amount of space occupied by something. This is called volume.

VOLUME IS THE AMOUNT OF SPACE OCCUPIED BY SOMETHING

An easier way to understand volume is to fill a glass. Whether you pour water, milk, juice, cement, or some other liquid into the glass, the amount needed to fill the glass is the same. That amount is the volume of the glass.

Here are some other examples of volume:
- The space in the water gun you fill up with water
- The space in Emma Jean's lunch box you filled with sand
- The space in the cat dish you fill up with food

Take a look at the milk in your refrigerator. It is probably either measured in a unit called a *gallon* or a *half gallon*. If you cut a half gallon in half, you have a 1/2 of a half gallon or a *quart*. Of course, all of this is much larger than all the measurements in our recipe.

No FAIR. THIS BOOK ISN'T SUPPOSED TO BE ABOUT FRACTIONS.

Okay. No fractions. How about this? There are four quarts in a gallon, and two quarts in a half gallon. Is that better?

Quarts are used to measure many of your favorite drinks. The label on a powdered drink mix tells you how many quarts of water to add. An envelope of lemonade mix will tell you how many envelopes to use for each quart of water. Sometimes bowls and pitchers even say on the bottom how many quarts they can hold.

If we divide a quart in two, you have a *pint*.

If you divide a pint in two, you at last come to a *cup*, as in our recipe. Cups get used often in cooking.

Divide a cup by eight, and you get one *fluid ounce*. Even though they are spelled the same, a fluid ounce is different from the ounces that are used to weigh small things.

If you divide a fluid ounce by two, you get a *tablespoon*. Hooray! Now we have another measurement for our recipe!

Tablespoons can be used to make tasty things. They are used to help make cakes, and pies, and cookies.

Tablespoons can also be used to bring us bitter things. They can be used to measure yucky sticky medicine.

And finally, at last, if you divide a tablespoon by three, you get a *teaspoon*.

That's a lot of measurements to remember. If you like charts, here's one to help you.

| 1 gallon | 2 half gallons |
| --- | --- |
| 1 half gallon | 2 quarts |
| 1 quart | 2 pints |
| 1 pint | 2 cups |
| 1 cup | 8 fluid ounces |
| 1 fluid ounce | 2 tablespoons |
| 1 tablespoon | 3 teaspoons |

Now, don't decide it's all too hard and you'll never cook anything. You can buy measuring cups, tablespoons, and teaspoons just for cooking. They are all marked with the sizes. They even have $3/4$, $2/3$, $1/2$, and $1/4$ cups, tablespoons, and teaspoons.

Part of the trick to understanding volume measurements is to use the right measurement at the right time. For example, suppose all recipes used only gallons for measurements. Our chocolate cake recipe would look like this.

2 Large eggs    1/128 gallon vanilla
1/8 gallon sugar    1/128 gallon soda
1/8 gallon flour
1/16 gallon cold water    1/32 gallon buttermilk
3/256 gallon cocoa
1/16 gallon margarine

Talk about fractions! Nobody would ever want to bake a cake again.

Okay. So fractions are way too hard. Let's measure everything in teaspoons.

> 2 large eggs
> 96 tsp. sugar
> 96 tsp. flour
> 48 tsp. cold water
> 9 tsp. cocoa

48 tsp. margarine
1 tsp. vanilla
1 tsp. soda
24 tsp. buttermilk

That's better than all those fractions, but you would still spend too much time measuring. You would have sugar and flour all over the kitchen. By the time you were finished, the mayor's fair would be over.

So when choosing a measurement, consider how large or small the amount to be measured is. Then pick the measurement that is best for the job.

Speaking of best for a job, you may have guessed that there is also a metric measurement for volume.

INTRODUCING ☆
THE
LITER FAMILY

• MILLI LITER
☆ • CENTI LITER ☆
• DECI LITER
• LITER
• KILO LITER

A quart and a liter represent almost the same amount of volume.

1 quart = 0.9464 liters (L)

Visiting the grocery store again, you will find that many of the larger soda pop bottles are measured in liters. Each bottle usually contains two liters of soda. Since one liter is almost the same as a quart, then each soda bottle contains a little less than two quarts, or half a gallon.

*Milliliters* (mL) are $1/1,000$ the size of a liter. To say it another way,

1 L = 1,000 mL

All the Milli members of the metric family have been very small. The same is true of the milliliter. To get an idea of how small, look at a can of soda. One can is usually twelve fluid ounces, which is about 355 milliliters. If you pour the soda from the can into 355 tiny little containers of the same size, each would have a volume of one milliliter.

Who uses such a tiny measurement? The same people who use the tiny milligram, scientists, doctors, and pharmacists, because they need very exact and tiny measurements.

Now take a look at the kiloliter, the big brother of the liter family.

1,000 liters = 1 kiloliter (kL)

That's more than 2,800 cans of soda.

But our recipe isn't measured in metric units of volume. It's measured in cups, ounces, teaspoons, and tablespoons. So let's see how Maverick and Elexus are getting along with the mayor's cake.

Looks like measuring a recipe is pretty important after all. Thanks to Maverick the mayor will have a cake for the contest.

A knock sounded at the door.

"Come in," the mayor said, wiping crumbs from his chin.

Baker Bob pushed the door open with his back while balancing a huge cake in the shape of City Hall in his arms.

"Wonderful!" The mayor exclaimed. "Baker Bob! What a fantastic job. Did you follow my grandmother's recipe?"

Baker Bob opened his mouth to explain he'd found it on his worktable beside a huge mess of batter. But before he could speak, the mayor grabbed the tray right out of Bob's hands.

"We need to get this to the judges right away," the mayor said, and he headed out the door with Baker Bob trailing behind.

# Chapter 8
# It's About Time

The mysterious hooded stranger couldn't believe his luck. Trusty Dusty and Mayor Marbles were leaving City Hall—together! He shrunk further behind the bushes as the pair passed by.

"You know how I hate these luncheons," the mayor grumbled.

"Now, sir," Trusty said, opening the car door so his boss could get in. "These volunteers have worked very hard on the city fair. They want to meet you."

"Okay. Okay," the mayor agreed. "But we can't take too much time. We need to figure out who this mysterious hooded stranger is and why he is causing so much chaos. And you still have to help me prepare for my TV interview. When is it again?"

"I'm not sure, but it's written on the calendar in your office."

"Better and better," thought the hooded stranger. Just the opportunity he'd been waiting for. When the mayor's car turned the corner toward the Niner Diner, the hooded stranger came out of hiding. Tucking his Zippo Zappo Eraser gun under his coat, he bound up the steps to City Hall. He dodged behind potted plants, ducked into doorways, and crept down the hall, until he finally stood outside the mayor's office. He studied the digital security lock on the door. It required an entry code. Numbers again. Oh, how he hated numbers. He glanced up and down the hall. No one was in sight. He slipped his gun out of his coat and adjusted the control dial.

Zap!

The security lock smoked, shimmered, and finally settled back into its place on the wall, numberless and useless.

The hooded figure slipped inside. A quiver of joy ran up his back as he closed the door. He was inside at last, standing alone in the mayor's office, his office, if his plan worked.

There were numbers everywhere. Big ones, little ones, fat ones, skinny ones! Once he became mayor, they'd be the second things to go.

"Right after Lostis Marbles and that Crusty Musty assistant of his," he muttered. Barely able to contain his joy, the hooded stranger dropped into the mayor's chair. With an evil cackle he pushed back his hood and smiled.

"I will destroy all numbers with my eraser gun. I will wipe out mathematics until chaos has come," he chanted as he twirled his Zippo Zappo Eraser gun like a gunslinger. "Now to put Phase Two of my plan into action."

Zap!

The numbers and hands on the clock vanished.

Zap!

The wall calendar disintegrated.

Zap! Zap!

A digital clock disappeared and the small bronze clock the mayor had received for his five years of service stopped ticking.

The hooded stranger, aka Acer Eraser, leaned back in the mayor's chair. "Perfect," he said. "Now the mayor won't have any idea when to attend his luncheons, or go home at the end of the day, and Trusty might forget to feed that mangy dog he calls Mongrel."

Acer's finger twitched on the Zippo Zappo Eraser. So many numbers. Too many numbers. With an effort, he restrained himself. He liked causing chaos, but all in good time. He'd zap this room clean after he became mayor.

Now he needed to finish what he'd come for. He opened the top drawer of the mayor's desk and found a pen. Removing the cap, he carefully added two small circles to a note on the mayor's calendar under that day's date.

Yes, that would do.

He glanced at his watch again. There was no time to lose. Now that he'd accomplished his mission, he had to hurry if he wanted to be in time. He'd leave the door ajar, just a bit, so Rusty Gusty What's-his-name would think he'd forgotten to shut it. Maybe they wouldn't notice the broken security system until later, after it was too late.

He slipped out of the room and was gone, leaving the mayor's office until another time.

Time is not only important to Mayor Marbles and Acer, but it has always been important throughout, well, time.

There is breakfast time, recess time, lunch and dinnertime, and the dreaded bedtime.

One of the oldest ways to tell time is by the position of the sun in the sky. So naturally some of the first time-keepers used the sun. It is called a sundial and is still used in some parts of the world today. By using shadows and the position of the sun, the day could be divided into parts, before noon and after noon.

Today, there are better ways to tell time than using the sun.

The kitchen oven timer, an hourglass, and wristwatches help people determine the time no matter where they are.

But the two timekeepers you see the most often are analog and digital clocks.

Analog clocks have hands, and, as you know, each hand has a different job. The shorter hand points out the hours and the longer hand tells how many minutes have passed.

To get a real feel of how long a minute is, put your head down on your desk. Have a friend watch the clock for you, and then raise your hand when you think it has been a minute. How close did you get?

Here are some more one-minute challenges.

How many times does your heart beat in a minute?

How many breaths do you take in a minute? (Breathing in and then out counts as one breath.)

How many push-ups can you do in a minute?

How many times can you say, "Peter Piper picked a peck of pickled peppers" in a minute?

Just as *minutes* divide an hour up into sixty parts, a *second* divides a minute up into sixty parts.

NO MONGREL. NOT THAT KIND OF SECONDS.

What can you do in a second?

When telling time with an analog clock, most people don't bother with the seconds. Seconds go by so quickly that by the time you do the math, and read the time, the seconds have changed. However, sometimes people need to know the exact hour, minute, and second.

Time including seconds would be written like this:

Hours   Minutes   Seconds

Now try reading a digital clock.

A digital clock has no hands. It tells the time in numbers by showing the time exactly the way you would write it.

So if it shows 4:35, you would say four thirty-five. Simple?

So which clock would you rather have, an analog or a digital?

Now, watch a clock for a whole day, not for every single second, but glance at it every now and then. At 8:00 you are at school. At 8:00 you might also be taking a bath or getting ready for bed.

Hold on! How can you be doing two different things at the same time?

Well, each day is twenty-four hours long, or the time it takes the earth to spin around once on its axis. (Remember that time was first measured by the sun.) A clock only shows twelve different hours. So in one day, a clock will repeat the same time twice.

So if a friend said, "Meet me at my house at 8 o'clock," when would you show up?

I MEANT 8:00 TONIGHT, NOT 8:00 THIS MORNING!

To stop this kind of confusion (and make friendships last longer), a little something extra was added after the time. The early morning hours from midnight to noon were called A.M., abbreviated from the Latin *ante meridiem*. The afternoon and evening hours from noon until midnight were called P.M., abbreviated from the Latin *post meridiem*.

So an agenda (that's a fancy word for schedule) of your day might look like this:

| | |
|---|---|
| 6:30 A.M. | Wake up |
| 6:40 A.M. | Dad yells at you because you're not up yet |
| 7:30 A.M. | Bus arrives at the bus stop |
| 7:35 A.M. | Mom drives you to school because you missed the bus |
| 8:00 A.M. | School starts |
| 12:00 P.M. | Lunch |
| 3:30 P.M. | Bus takes you home |
| 6:00 P.M. | Dinner |
| 7:00 P.M. | Bath |
| 8:30 P.M. | First bedtime |
| 9:00 P.M. | Second bedtime |
| 9:30 P.M. | Really-in-trouble bedtime |
| 11:59 P.M. | Reading by flashlight under the covers |
| 12:00 P.M. | Dad says, "Are you asleep in there?" |
| 12:01 P.M. | Flashlight off |

Notice that 12:00 P.M. refers to noon and 12:00 A.M. refers to midnight. If you're still up at 12:00 A.M. and get caught, you'll really be in trouble.

Many times it is important to know what time something will be over based on how long the event is going to last. For example, suppose you are invited to a birthday party. Your friend tells you the party starts at 1:00 P.M. and will last for three hours. What time should you tell your parents to pick you up?

1:00 P.M. + 3 hours = 4:00 P.M.

That's not too hard. However, suppose the party starts at 11:00 A.M. and lasts for three hours? Now what time should your parents come?

11:00 A.M. + 3 hours = 14:00 A.M.

Hold it! There is no such thing as 14:00 A.M. You'll need to think it through this way:

a.  11:00 A.M. + 1 hour = 12:00 P.M. (noon)
b.  There are still two more hours of party left.
c.  That means your parents should pick you up two hours past noon or 2:00 P.M.
d.  Since you went past noon into the afternoon, the label for the time switched from A.M. (morning) to P.M. (afternoon).

Figuring out how much time has passed may also require you to consider the minutes. For example, suppose your school begins at 8:45 A.M. and ends at 3:15 P.M.

a.  From 8:45 A.M. to the nearest hour, 9:00 A.M., would be fifteen minutes.
b.  From 9:00 A.M. to noon would be three hours.
c.  From noon to 3:00 P.M. would be another three hours.
d.  From 3:00 P.M. to 3:15 P.M. would be another fifteen minutes.
e.  Add the hours together: 3 + 3 = 6 hours.
f.  Add the minutes together: 15 + 15 = 30 minutes.
g.  So, you spend six hours and thirty minutes in school each day.

Decide how much time has passed from the starting time to the ending time on each of these.

1.  You rode the first ride at an amusement park at 10:15 A.M. and the last ride at 5:45 P.M. How much time passed between the first ride and the last ride?
2.  You attended a roller skating party that started at 4:30 P.M., and you skated until 7:15 P.M. How long did you skate?

3. You went to a movie that started at 1:55 P.M. and ended at 3:20 P.M. How long was the movie?

*3. 1 hour and 25 minutes*
*1. 7 hours and 30 minutes; 2. 2 hours and 45 minutes;*

This A.M. and P.M. stuff really worried the government officials like Mayor Marbles. When they needed something done, everyone absolutely had to get the time right.

Instead of using the A.M. and P.M., they came up with a method called military time. For A.M. hours, they used the same time as everyone else. So, 6:00 A.M. becomes 06:00 in military time. The difference between the two types of time is the afternoon and evening hours. Instead of starting over with 1:00, 2:00, or 3:00, military time keeps on counting with 13:00, 14:00, and 15:00. So when you are at lunch and the clock says it is 12:59 P.M., when one more minute ticks off, it will be 13:00 in military time, instead of it becoming 1:00.

Notice that the difference between the 1:00 and 13:00 is twelve. So to change military time to P.M. time, just subtract 12 from the military time.

Example:

15:27 = 15 – 12 = 3:27 P.M.

To change P.M. time to military time, add twelve hours.

Example:

5:40 P.M. = 5 + 12 = 17:40

The 12 o'clock hours are the exceptions to this rule.

12:00 A.M. (midnight) is 00:00 in military time.
12:00 P.M. (noon) is 12:00 in military time.

# Time Pattern Puzzle

Now it's time for a little practice. First, match the times at the left to the military time and their associated puzzle piece at the right.

A1 3:30 A.M.

A2 12:00 P.M.  0:00  2:00

A3 5:00 P.M.

A4 2:00 A.M.  3:30  4:30

B1 4:30 P.M.

B2 7:30 P.M.  5:00  12:00

B3 11:30 P.M.

B4 8:00 P.M.  13:00  13:30

C1 12:00 A.M.

C2 5:00 A.M.  14:00  15:00

C3 11:00 P.M.

C4 9:00 P.M.  16:30  17:00

D1 10:30 P.M.

D2 2:00 P.M.  18:00  19:00

D3 4:30 A.M.

D4 1:00 P.M.  19:30  20:00

E1 1:30 P.M.

E2 3:00 P.M.  21:00  22:30

E3 6:00 P.M.

E4 7:00 P.M.  23:00  23:30

Second, color in the squares below with the picture next to your answers and discover a secret message. Check your answer at the end of the chapter.

|   | A | B | C | D | E |
|---|---|---|---|---|---|
| 1 |   |   |   |   |   |
| 2 |   |   |   |   |   |
| 3 |   |   |   |   |   |
| 4 |   |   |   |   |   |

Now that your twenty-four hour day is all settled, how about the rest of your life?

Sticking with the earth-sun-moon way of telling time, a *month* was the time from full moon to full moon. A *year* was the time it took for the earth to go around the sun once.

The earliest calendars had ten months. Eventually two more months were added to make twelve. The months were named after Roman deities, emperors, or just the number of the month. We still use these names today. Try matching our months to their Roman origins. (Hint: Remember that there were originally only ten months in a year.)

|  Month | Origin |
|---|---|
| 1. January | A. *Aprilis*, to open |
| 2. February | B. *Augustus*, Roman emperor |
| 3. March | C. *decem*, ten |
| 4. April | D. Februss, old-Italian god, or else |
| 5. May | from *februa*, signifying the |
| 6. June | festivals of purification |
| 7. July | celebrated in Rome during |
| 8. August | this month |
| 9. September | E. *Janus*, Roman god of beginnings |
| 10. October | F. Julius Caesar, Roman ruler |
| 11. November | G. *Juno*, Roman queen of the gods |
| 12. December | H. *Maiesta*, the Roman goddess of |
|  | honor and reverence |
|  | I. *Mars*, Roman god of war |
|  | J. *novem*, nine |
|  | K. *octo*, eight |
|  | L. *septem*, seven |

1.E 2.D 3.I 4.A 5.H 6.G 7.F 8.B 9.L 10.K 11.J 12.C

All the Roman months had thirty days, so 30 times 12 made 360 days in their year.

Unfortunately, the earth refused to cooperate with this simple math. It insisted on taking 365 1/4 days to go around the sun. In the past, this problem was solved by adding an extra month to their calendar every so often.

Today, the length of a month varies from twenty-eight days to thirty-one days. This verse may help you remember which month is what length.

THIRTY DAYS HATH SEPTEMBER,
APRIL, JUNE, AND NOVEMBER, AND
ALL THE REST HAVE THIRTY-ONE
EXCEPT FEBRUARY HAVING 28.
BUT LEAP YEAR COMING ONCE
IN FOUR, FEBRUARY THEN
HATH ONE DAY MORE.

Leap Year? What's that?

The earth actually takes 365¼ days to go around the sun, so every fourth year, February gets another day to make the years come out even. To tell if a year is a leap year, divide it by four. If it has no remainder, then that year is a leap year.

Example 1

   $2006 \div 4 = 501$ with a remainder of 1
   Not a leap year
   Days in February = 28

Example 2

   $2008 \div 4 = 502$ with no remainder
   A leap year
   Days in February = 29

What year were you born? Was it a leap year?

If you were born on the last day of February in 2000 (a leap year), does that mean you only have a birthday every four years? If so, how old will you be in 2020? Would you be twenty years old or only five?

Now that you know about time, here's a challenge.

How old are you in years?

How old are you in months? (Grab a calculator for this one.)

How old are you in days? (Hint: Don't forget to add an extra day for every four years.)

How old are you in hours?

How old are you in minutes?

How old are you in seconds?

Are you feeling old, yet?

That ends this lesson in the nick of time because the mayor's luncheon is over and Trusty and Mayor Marbles have returned to their office.

"Are you sure your interview is at eight o'clock, sir?" Trusty asked. "I thought you were appearing on the Good Afternoon Mathopolis' talk show."

"That's the time I have written on my calendar," Mayor Marbles said. "See? It says eight o'clock. You wrote it yourself."

"I know, sir, but—"

"Mayor! Mayor!" Belinda, the bookkeeper, burst into the room. "Quick, turn on the TV."

Trusty picked up the remote control and punched the power switch.

"I dislike speaking ill of anyone," an overly sweet voice crooned. "But what Mathopolis needs is a change in leadership." The picture faded in, and there, in his silver suit, with his silver hair immaculately combed, standing exactly where Mayor Marbles should have been, was Acer Eraser on the 3:00 "Good Afternoon Mathopolis" talk show.

Trusty slapped his head. "You missed the interview!"

"Impossible," exclaimed the mayor. "My desk calendar says—"

"This three has been changed to an eight," interrupted Trusty examining the calendar.

"But who? Why?" asked the mayor, sagging into his chair.

Trusty looked around the room. "These clocks have been tampered with. Someone wanted you to miss that interview. It had to be someone who hates numbers as much as you love them."

"That guy who's been hanging around town zapping everything hates numbers," said the mayor. "Maybe it was him."

"But if it was him, how did Acer Eraser know you would miss the interview?" Suddenly Trusty knew the truth. "Unless Acer is the mysterious stranger."

Trusty pointed at the television.

"And if you elect me mayor," Acer was saying. "I'll see to it that the mysterious hooded stranger never zaps another number. Mathopolis will be restored to its former glory. Your city will be chaos free and safe from the threat of that evil no-numbers fiend."

"He wants to take over," said Trusty. "He wants your job."

"The people will never allow it," insisted the mayor.

Just then laughter and applause erupted from the television. The audience liked Acer. Even Mary, the host, was laughing and smiling at him.

"We've got to stop him," said Trusty.

The mayor nodded. "But what can we do?"

Answer to Military Time Puzzle

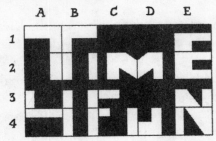

# Chapter 9
# Temperature

"What are we going to do?" Maverick said, surveying the scene in front of him.

"We have to prove that Acer is behind this chaos," Elexus said.

Mongrel barked his approval.

"We've got to find him first," said Maverick.

"You start here and work your way north," said Elexus. "I'll take the south side of town."

Maverick nodded. "If you find him, send up a lightning bolt signal. You watch for my beamed beacon. Otherwise we'll meet back at City Hall in three hours."

Elexus nodded. "We better get going before his trail gets cold." She took a mighty leap and disappeared into the sky.

Ever since we've had hot and cold, we've needed to be able to measure it.

Air temperature tells you whether to wear a coat or shorts when you go outside. Your body temperature tells you whether you are sick or well. You set the oven to a certain temperature in order to cook your food. Some recipes say your dinner is ready when the food reaches a certain temperature.

Temperature tells us how hot or cold something is. Scientifically, temperature is how fast or slow the microscopic parts (molecules) of an object are moving. The faster the molecules move, the warmer the object becomes. When the molecules slow down, the object becomes colder.

Think of it this way. Suppose you start running around and around your room at home. The more you run, the more energy you use, and the hotter you become. When you finally flop onto your bed to rest, you start using less energy and you cool off.

In the same way, temperature measures how much energy molecules are using.

To find out the exact temperature of something, you need a scale of some kind. For height, you used either feet or meters. For weight, you used either pounds or kilograms.

Temperature is measured in *degrees*. The symbol for degrees is a small circle. For example, eighty-five degrees would be written as 85°. There are three temperature scales that are used today.

1. *Kelvin (K) scale*—a metric system measurement used by scientists and for astronomical temperatures. No degree circle is used after the K in the Kelvin scale.
2. *Celsius (°C) scale*—used in most of the world to measure air temperatures.
3. *Fahrenheit (°F) scale*—used in the United States to measure temperatures.

For each of these methods of measuring, zero means something different.

In Kelvin, zero is the temperature at which the smallest particles (molecules) would stop if they had become too cold to move. This is called *absolute zero*. Since this could never happen in real life (at least let's hope not), zero on the Kelvin scale has never been reached.

In Celsius, zero is the temperature at which water freezes.

In Fahrenheit, zero is the coldest temperature that scientist Gabriel Daniel Fahrenheit could create with a mixture of ice and ordinary salt.

Using each of these scales, here are some temperatures you might be familiar with.

|  | Fahrenheit | Celsius | Kelvin |
|---|---|---|---|
| Frozen water | 32° | 0° | 273 |
| Boiling water | 212° | 100° | 373 |
| The human body | 98.6°* | 36.8° | 309.8 |
| Cool enough for a sweater | 59° | 15° | 288 |
| Warm enough to swim | 80° | 35° | 308 |

*Some modern research has indicated that 98.2° Fahrenheit may be a more accurate figure.

Since most of the world outside the United States uses Celsius to measure temperature, it would be handy to be able to change Fahrenheit degrees to Celsius.

Here is the formula.

$$\text{Celsius} = \frac{5}{9} \times (\text{Fahrenheit} - 32)$$

Yes, you have to use fractions, but they really aren't so bad.

Suppose you know that it is 77°F and you need to change that to Celsius. Start by subtracting 32 from 77. (You begin with the subtraction because it is in parenthesis, just like this sentence. You always do what is inside the parenthesis first.)

$$77 - 32 = 45$$

Now you want to multiply that number times the fraction $\frac{5}{9}$.

So, multiply 45 times 5 (the top number of the fraction).

$$45 \times 5 = 225$$

Then divide the answer by 9 (the bottom number of the fraction).

$$225 \div 9 = 25$$
$$77°F = 25°C$$

To change Celsius to Fahrenheit, just reverse what you did up above. The reversed formula would look like this.

$$\text{Fahrenheit} = \frac{9}{5} \times \text{Celsius} + 32$$

Notice that to reverse the $\frac{5}{9}$, you will be multiplying times the reciprocal (turned over version) of $\frac{5}{9}$, which is the fraction $\frac{9}{5}$. And where you started by subtracting 32, you will end by adding 32.

Now you try it.

Bradford Bear has invited Mongrel to come for a visit. He told Mongrel that during the day it is about 45°C. What should Mongrel pack to wear?

To help Mongrel, you will need to change 45°C to Fahrenheit.

Start by multiplying 45 times $^9/_5$.

Multiply 45 times 9 (the top number).

$$45 \times 9 = 405$$

Divide that answer by 5 (the bottom number).

$$405 \div 5 = 81$$

Now add 32 to your answer.

$$81 + 32 = 113$$

45°C = 113°F.

As long as we are changing things around, let's go ahead and change to the Kelvin scale. No fractions are needed for this formula. Just take the Celsius temperature and add 273.

So, 45°C = 45 + 273 = 318 K

Check what you know so far. Add a K, °F, or °C (we're not talking Colonel Sanders, here) after each of the following to show the best temperature scale for drinking each without burning or freezing you tongue.

1. Hot chocolate:  40___
2. Lemonade:  280___
3. Milk:  55___
4. Coffee:  313___
5. Soda pop:  15___
6. Iced tea:  40___
7. Hot tea:  100___

1. °C  2. K  3. °F  4. K  5. °C  6. °F  7. °F

As the temperature increases or decreases, certain properties of a substance will change as well. For example, water changes according to how hot or cold it is. When water reaches a temperature of 32°F, it becomes ice, a solid.

When the frozen water warms up, it becomes a liquid.

If you heat the water to a temperature of 212°F, it boils and gives off steam. Steam is a gas.

To find out how hot or cold something is, you use a thermometer. (*Thermo* means "heat;" *meter*, as used here, means "to measure;" *thermo* + *meter* = thermometer.)

There are several different kinds of thermometers.

The most common is the bulb thermometer. It is a sealed long glass tube with a bulb at the end with liquid inside. You or your parents may have had one stuck under the tongue to take a temperature when sick. There might even be one on your back porch to show the outside temperature.

Bulb thermometers work using the idea that when a liquid is hot it takes up more space than when it is cold. When the liquid inside the thermometer is cold, most of it shrinks up and goes down into the bulb. When it gets warmer it expands and moves up into the glass tube.

A scale on the side of the tube measures the temperature based on how far the liquid expands (gets bigger) or contracts (gets smaller).

Most bulb thermometers use a liquid called mercury. Mercury doesn't freeze until –38°F, and it doesn't boil until 673°F. So mercury in a thermometer allows it to measure much higher and lower temperatures than water. It can measure the weather, volcanoes, and maybe how hot lightning gets.

Acer Eraser knew that sound. Elexus had spotted him. Straddling his Supersonic Scooter, he frantically looked around. He needed extreme chaos in order to escape. He aimed his Zippo Zappo Eraser gun and fired.

Kazam!

The lock on a passing truck sprung open. Ice spilled from the inside. Cars careened, swerved, and skidded. Other motorists squealed to a stop to avoid hitting them. Children clutched their parents' hands as they slipped and slid on the ice.

Acer cackled and sped away down an alley. That should keep Elexus out of his hair for a while.

# Chapter 10
# Guess What?

"Whoa," Acer Eraser cried as he lost his balance and slid halfway down the roof of City Hall. He wanted to hide behind the big numeral six that stood on top of the building. He dug his heels into the tiles and hooked his arm through the six to stop his fall; then he scrambled into a seated position.

"I will destroy all numbers with my eraser gun. I will wipe out mathematics until chaos has come," Acer mumbled to himself looking around. There had to be something he could zap to increase the chaos today.

As if adding an exclamation point to these words, a huge crane appeared, bouncing and rumbling down the street, leaving a wake of black smoke behind. The poor driver was holding on for dear life.

At the sound, the mayor hurried out of the Niner Diner with Trusty trailing behind him.

"He needs to know now, sir," Trusty was saying.

"Who needs to know what?" the mayor asked, waving to the people.

Trusty waited for Mayor Marbles to stop before he went on. "The crane driver needs to know how much the fair sign weighs." He pointed down the street to where the crane had stopped.

"Well, why are you asking me?" Mayor Marbles asked. "How should I know?"

"You ordered the sign, sir," Trusty Dusty replied. "The weight must have been on the receipt."

"I would guess it was maybe fifty pounds."

"No, no, no," Trusty said. "The driver needs the correct information. If the sign is too heavy, he might break the crane."

"Tell you what," Mayor Marbles replied. He pulled a handful of wadded papers out of his pocket and shoved them at Trusty. "I'm sure the receipt is in here somewhere. You keep looking for it. I need to find some babies to kiss." The mayor wandered down the street shaking hands that didn't want to be shook, kissing babies, and smiling.

Trusty rummaged through the rumpled receipts in his hands. "Fifty trophies, hay for the camels."

The crane operator frowned and impatiently drummed his fingers on the steering wheel. "Is this going to take all day?" he muttered.

Acer shouldered his gun.

"Wait. Here it is. It says—"

Kazowie!

"Where did the numbers go?" Trusty looked around quickly, and then back at the useless receipt in his hand. "Now what?"

Outside the Niner Diner, Mongrel set aside his breakfast bone, threw back his head, and sent a deep howl into the morning air.

"You called?" Elexus asked.

In a good imitation of a bird dog, Mongrel pointed toward the roof.

"Not again!" Maverick said, looking up just in time to see Acer take aim, and fire toward the crane.

Faster than lightning, Maverick whipped a saucer off his belt and threw it into the air.

Sparks flew as the beam reflected off the disk and bounced back toward Acer Eraser.

"Let's get him," said Elexus. "I'll fly. Maverick, you take the fire escape."

Mongrel's frantic barking stopped them both in their tracks.

"What's wrong?" cried Maverick.

"Trusty! Quick!" said Elexus. "How many men did it take to carry the sign?"

"Eight," he replied.

"That's all I needed to know," she said flying to the rescue. "Stop," she called to the operator. "You need a larger crane."

The crane operator frantically worked the controls, and lowered the sign back to the ground.

Elexus flew back to join her friends.

"How did you know?" Maverick asked.

"Volume, time, height, and weight, the answer is to estimate," Elexus said with a smile.

Estimation saved the city's fair sign, and it can save you from all sorts of problems, too. But just what is this nifty skill, and how do you know when to use it?

Say you asked three kids at random to tell you something they'd like to estimate.

All three of Mongrel's subjects have something they need to know, and their questions all involve different types of estimating. But how do they know they need to estimate?

Unfortunately, most problems don't come with signs. The first boy finds out he needs to estimate the hard way. Let's watch and see.

His question was, "How much does the lunch lady weigh?"

1. He could ask.

2. He could try to measure.

3. He could guess.

4. He could estimate.

It isn't hard to figure out that estimating is the best solution. But what exactly is estimating?

Sometimes when you are trying to understand something, it helps to know what it is not. For example, just because your little sister wraps a cape around her shoulders doesn't mean she's a superhero. Thank goodness!

So what isn't estimating? Well, it isn't precise. If you estimate the weight of your lunch lady, you will not get the exact answer unless you are very lucky. If you estimate how long it will take you to earn enough for a Burping Baby, it will not be exact either. If you want the exact answer to how many gum balls are in the jar at the library, you need to count them.

So, it is safe to say that estimating is not precise.

And, even though you may have had the experience of estimating the exact number of gum balls in a jar and winning a prize, this is not the job of estimation either. The job of estimation is to come as close as you can given the tools—or lack of tools—you have, the time you have, and the knowledge you have.

In fact, the purpose of estimating is to be *wrong*. Yes, *wrong*. And being wrong can help you win prizes, and being wrong can make you look smart. Wow! That's a double hitter!

On the other end, estimation isn't just a guess. If someone says to you, "I'm thinking of a number between 1 and 10. Guess which one it is." Well, that's what you do. You guess. You have no idea what number another person is thinking, unless you have crazy X-ray vision. You also have no way of figuring it out, even if you try. You might as well pick a number out of a hat.

But if the same person said, "Guess how tall I am," you'd think about it. That brings us to what estimating is.

Estimation is in the middle. It's between the precision of measuring, calculating, or counting and the fun of guessing. Estimating is an educated guess. It's using your mind and other information to quickly figure out a reasonable answer. You'll learn about this in the next several chapters.

But how did Elexus so quickly judge the weight of the sign? Although he wouldn't admit it to her for the world, Maverick is wondering the same thing.

"Well," Elexus explained to Mongrel as Maverick listened in from a few feet away. "I knew that one man could carry about forty pounds. So I multiplied the eight men times the forty pounds. That meant together they could carry about 320 pounds."

"I also know that this crane can only handle about 250 pounds because it said so on the side," Elexus went on. "So the sign was going to fall. Estimating isn't guesstimating. It really does work."

"Height, weight, volume, and time, if it's not precise, it's a crime," Maverick countered.

Mongrel shook his head. Superheroes could sure be stubborn sometimes.

# Chapter 11
# Estimating 101

"Whew! That was close," Maverick said. "Too close."

He was in the park at the small clearing by the pond, kneeling down besides his tool belt.

"True," Elexus replied from her position in a low tree branch. "If that sign had fallen, someone might have really been hurt."

Maverick continued spreading out his tools, carefully laying each of them on the ground. "I'll need another reflector disk and a new tape measure. This needs to be recharged, and . . ."

"We have to come up with a plan to stop Acer. Why are you wasting your time with these things?" Elexus interrupted, jumping off her limb.

"Height, weight, volume, and time," Maverick mumbled. "If it's not precise, it's a crime."

Elexus picked up the lasso and lazily twirled it overhead. "Volume, time, height, and weight, the answer is to estimate."

"Hey, be careful with that," said Maverick.

She looped the rope around her arm and flew up into the air.

"Come back here," Maverick yelled, pointing to the ground beside him.

"You want this?" she taunted as she hovered high over his head twirling the rope in ever widening circles.

"Elexus, don't!"

"Don't what?" She let the rope fly, and within seconds Maverick was tied.

Laughing, Elexus flicked the end of the rope she still held. "Will you admit that estimating is better than measuring?"

"Never!" Maverick said, squirming against the ropes.

"Then I'll see you." Elexus started to take off, but a familiar bark sounded close by.

Elexus let go and Maverick squirmed free just before the dog arrived.

The dog looked at Elexus, then Maverick, and finally at the rope piled at Maverick's feet.

"I was practicing," Maverick mumbled as he gathered the lasso.

Mongrel woofed twice, and then trotted off toward the fair grounds. Without a word, Elexus helped Maverick get his tools back on his belt, and then they followed.

Elexus would be pleased to know, we are still talking about estimation. In the last chapter, we learned estimating is not about being precise, and it isn't a guess either. Rather, as Elexus knows, estimating is somewhere in between, a pickle-in-the-middle.

So where does Elexus start? She's certainly not going to pull a number out of her hat, and she's not going to get out measuring tools and measure. That's Maverick's job. But she does know to start by thinking about how big or small an answer will be. First, she wants to get the right size number.

For example, when you are estimating a calculation like $13 \times 11$, the answer could be in the 1's or 10's or 100's.

Bigness and smallness comes in all sorts of packages. For estimating the measurement of length, you start with the question of whether it's inches, or feet, or yards, or miles.

## Which is most likely true of this bone?

It is 9 inches long.

It is 9 feet long.

It is 9 yards long.

PAUSE FOR THOUGHT

Although Mongrel may wish his bone is nine yards or nine feet long, its width is closer to the width of his dog bowl, about nine inches.

It's simple. For estimating weight, decide whether to use ounces, pounds, or tons. For time, decide on seconds, minutes, hours, days, weeks, or months. For volume, choose teaspoons, cups, pints, or gallons.

Just remember that deciding bigness or smallness is the place to start with estimating. Often you don't even realize you are doing it because it is a natural way to think. How fast can you choose the size in the situations below?

You are handed a glass of milk. How much milk is in the glass? Is it small like the teaspoon? Is it medium sized like a cup or pint, or would you guess it would be a gallon?

If you are a giant, it would be a gallon. If you are a mouse, it would be a teaspoon, but for everyone else it would be medium sized, cups or pints.

A fast runner is racing a hundred-meter dash. How long would it take him to cross the finish line? Would it be measured in seconds, minutes, hours, or days?

If size is the place to start, comparing is the heart and soul of estimating. In fact, when you estimate, you are up to your ears in comparing.

To estimate measurements, you compare length, weight, volume, time, temperature, and amounts. To estimate calculations, you also compare. Luckily, you are an expert in comparing, whether you realize it or not.

Now you need to apply your expert comparing skills to estimating. It's all about comparing what you don't know to what you do.

For example, you don't know how tall the man on stilts is, but you do know how tall your dad is. You don't know how many hot dogs you can eat today, but you do know how many hot dogs you ate yesterday. You don't know what $24 \times 4$ equals, but you do know that $25 \times 4 = 100$ (because you know that four quarters make a dollar). You don't know how many gum balls are in the jar, but you can count how many are in a small section of the jar.

First, let's review some key words and symbols.

So the key concepts for comparing are *greater than, less than,* or *equal to*.

The height of the man on stilts is greater than (>) the height of your dad. Your hunger yesterday is equal to (=) your hunger today. Lastly, $24 \times 4$ is less than (<) $25 \times 4$.

Sometimes when you are comparing, you compare in your head. For example, you can compare your hunger today with yesterday's, or you can compare 24 × 4 with 25 × 4. How? You compare how two or more things are the same and how they are different. But sometimes you compare things and people by putting them next to each other.

You can do this with a friend. Stand side by side. Who is taller? If you are forty inches tall, stand next to him and figure out his height. Is it greater than your forty inches? Is it less than? Or is his height equal to yours?

How long is that hot dog? Is it really twelve inches long? You probably didn't carry a ruler to the restaurant, but what could help you figure it out? If you know the length of your foot you can compare it to that, but don't eat the hot dog if you compare them side by side!

Is the hot dog equal to twelve thumbs?

Finally, sometimes when you compare, you end up with a range.

Take the man on stilts. He is taller than your dad. Your dad is six foot. But he is shorter than the tent opening. The tent opening is 7'1". That means the man on stilts is between 6' and 7'1". You can also use a range to determine the height of a friend.

HOME, HOME ON THE RANGE, WHERE THE DIGITS AND DECIMALS PLAY...

You can also end up with a range when you are estimating calculations.

Given the question, "What is 11.2 × 11.9," you can use a comparison to get a range for the answer.

What do you know?

Well, 10 × 10 is 100. The answer must be greater than that because both 11.2 and 11.9 are greater than 10.

And you know that it's less than 12 × 12 and 12 × 10 = 120 and 2 × 12 = 24 so 12 × 12 = 120 + 24 = 144. (The answer must be less than 144 because both 11.2 and 11.9 are less than 12.)

So, 11.2 × 11.9 is greater than 100 and less than 144 (also written as 100 < the answer < 144).

Now you try the following problems:

1. Given the problem 23 × 5. Compare it to another problem you know the answer to.
2. Compare a stick of gum to something you know the length of.
3. Give a range for how tall you are.

3. one yard < my height < my mom's height of 5 feet 10 inches.
2. stick of gum > my thumb = about 1 inch
1. 23 × 5 < 25 × 5 = 125 (think quarters)

Okay, to begin estimating, first figure out the size of the answer, and then compare. This is just the beginning, though. There are a few more things to learn because different ways to estimate are used for calculations, measurements, or doing something called sampling.

These will all be explained in the next few chapters, but for now, Maverick, Elexus, and Mongrel have arrived at the fair grounds.

Mongrel led Elexus and Maverick to the main gate.

"What's the matter?" Maverick asked.

"Thank goodness you're here," said Tim the ticket taker. "Last year's attendance numbers have vanished."

"Acer again," said Elexus.

Maverick shrugged. It didn't seem like much of an emergency to him. "Why do you need those?" he asked.

"We use them to help us figure out how many people might be coming to the fair tomorrow."

"Well, count them when they get here," said Maverick. "That's exactly how I'd do it."

"We can't wait that long. We need to plan ahead so we'll know how many tickets to print. Not to mention buying food and ice, building the stands, ordering outhouses—"

Elexus pushed Maverick aside. "Volume, time, height, and weight. The answer is to estimate."

Maverick folded his arms and glared at her.

"It's simply a question of size. Did more than a hundred people come last year?" she asked Tim.

"Yes."

"More than 1,000?"

"Yes."

"More than 10,000?"

Tim hesitated and scratched his head. "I don't think so."

"Excellent," said Elexus. "I would estimate that for the fair you will need supplies for less than 10,000 people."

Maverick harrumphed.

Elexus glared at him, then continued. "Since the fair lasts for five days, you should buy only enough for the first day's attendance. That would be about 2,000 people. After that you should be able to see if your estimate is too high or too low and adjust for the next four days."

Tim's face broke into a wide smile. He reached out and pumped Elexus's hand so hard her teeth chattered. "Thank you. You've saved the day."

Elexus put her hands on her hips and shot Maverick a triumphant smile. "Volume, time, height, and weight. The answer is to estimate."

Now you know how to estimate...turn the page and learn how to truncate!

# Chapter 12
# Truncating

Acer peered out from under his hood. Wiping out entire numbers was getting boring. Today, he'd try something new, and wipe out only part of a number to confuse people. The fair was the perfect place to start.

Acer rubbed his hands together in delight. "I will destroy all numbers with my Zippo Zappo Eraser gun," he said softly. "I will wipe out mathematics until chaos has come." He zoomed away from the fair grounds on his scooter. It was a shame he couldn't hang around to watch more of the fun, but Elexus and Maverick had gotten close several times. Too close.

Still, his plan was unfolding nicely. People were getting tired of the chaos. His campaign for mayor was going well. And everyone still thought it was the mysterious hooded stranger ruining the day. After he won the election, people would know soon enough who really was in charge. He hadn't done any calculations, but by his reckoning, he had this election in the bag. When he was done with this campaign, not only would he be mayor, but no one in Mathopolis would ever do calculations again.

But, is Acer right? Just because a few calculators have been Zippo Zappoed, does that mean no one in Mathopolis is calculating?

Oh my. It's easy to see why someone might need to know how to estimate calculations like +, −, ×, and ÷. You never know when a real calculator-eating horse may gallop by. But when else would you want to estimate calculations?

1. Sometimes you don't need exact answers.
2. It can be helpful to estimate even if you have done the calculation—just in case you have messed up.

AND IN CASE YOUR HORSE
EATS YOUR CALCULATOR...

3. Or you could be stranded on a desert island without a calculator.

Now, let's get going on estimating calculations. Pretend that your mind is a Mighty Number Transformation Machine. Can you see it?

MIGHTY NUMBER
TRANSFORMATION MACHINE

You don't have to pretend because your mind actually is a Mighty Number Transformation Machine. When you estimate, it can input any number, transform it, and output another number.

So just how does your mind do this?

One way is to shorten the numbers. In math this is called . . .

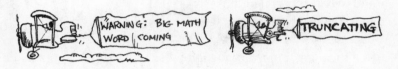

It's a big word, but it's mighty helpful. Mongrel will demonstrate how a dog truncates the number 818,564.89.

So truncating in dog language is biting off some digits. It's simple and tasty. Notice also that Mongrel wasn't too greedy; he didn't bite off the first eight. If he had, there wouldn't have been any number left.

The idea behind biting off a number is to get smaller numbers that are easier to work with when you add, subtract, multiply, and divide. Most of the time, you take a big enough bite to keep the digit furthest to the left.

789 becomes 7
812 becomes 8

And most people can easily add 7 + 8 either in their heads, or by counting on the fingers and toes.

Well, most people can.

When you are chewing up digits, you have to keep in mind what the remaining digits mean. Remember in Chapter 2 it said that every place has a meaning and every meaning has a place? It's important when you bite off part of a number to not think that 812 became just plain eight. There's a lot of difference between eight maids a milking and 800 maids a milking.

Keeping in mind what the remaining digits mean will help you when you add or subtract two numbers of different lengths. You can't just bite whatever you want and then add them up.

Let's look at this example:

$$453{,}432$$
$$-\ 34{,}678$$

Let the truncating begin.

What did he do wrong? Did he throw a spitball when no one was looking? Was someone trying to trip him, and the teacher wanted to stop his fall? Or was his answer wrong? And if so, what's wrong with it?

The problem is he didn't take the same size bite from each number. Remember, the four stands for 400,000 and the three only stands for 30,000 so the four and the three aren't in equal places. When digits aren't in equal places, you can't treat them the same.

For addition and subtraction you have to take the same size bite from each number to make them equal. (Multiplication and division can have bites that are different sizes.)

Let's ask Mongrel to explain. He has a big enough mouth to show us how this works. No offense, Mongrel.

From Mongrel's greedy bite, you can learn that since you need to take the same size bite from both numbers, when adding and subtracting, you better bite the smaller number first. If you aren't a dog and aren't biting, you would say, "truncate the smallest number first, and then truncate the other numbers the same amount of digits."

Let's let Mongrel take another try. Here's Mongrel's truncating, take two.

See? Mongrel is left with 45 − 3 = 42. But this is only part of the answer. He took off places, so how does he put them back?

And for that answer, here comes Mongrel with a donut machine. This is the tastiest part of truncating.

Remember, Mongrel bit off four digits from each number. It makes sense that you would put those back. But you don't put them back as the digits you took off. You do put them back as sweet donuts (also called zeros). But how many zeros do you put back? The following rules will help.

### The Biting Donut Table

| Operation | Number of Donuts to Use |
| --- | --- |
| Adding and Subtracting | Count the number of places bitten off from the smallest number, and put that number back as donuts. |
| Multiplying | Add the number of places bitten off, and put them back as donuts. |
| Dividing | Subtract the smallest number of places bitten from the largest number of places bitten off, and put that number back as donuts. |

Since 453,432 − 34,678 is a subtraction problem, you put back the places bitten off from the smallest number, so when four digits are bitten off, four zeros go back. The four zeros put the forty-two in its proper place of 420,000.

As you read earlier, multiplication and division are different. With multiplication and division, you don't have to truncate or bite off the same amount of digits because you simply add up the places taken off of both numbers and put them all back as zeros.

 Try the problem 2,111 × 25. What would you bite off? Remember the bites don't have to be the same size with multiplication.

One way is 2,111 becomes 2 and 25 stays as 25. 2 × 25 = 50. Now add back zeros. The first number lost three digits and the second number lost zero digits; 3 + 0 = 3. Therefore, 50 with three more zeros becomes 50,000.
Here's another way: 2,111 becomes 2 and 25 becomes 2. 2 × 2 = 4. The first number still lost three digits. The second number lost one digit; 3 + 1 = 4 digits. Therefore, the answer 4 with four zeros put back becomes 40,000.

The second answer isn't as close as the first estimate because more was truncated or bitten off, but it is still an estimate.

Here are a few more problems:

1. 780 + 912
2. 1,506 − 427
3. 84,404 ÷ 215

1. 7 + 9 = 16 and put back two zeros to get 1,600.
2. 15 − 4 = 11 and put back two zeros to get 1,100.
3. 844 ÷ 2 = 422. Since each number lost the same amount of places, you don't need to put back any zeros. This could also be done in other ways, like 84 ÷ 2 = 42 and put back one zero to make it 420.

This chapter on truncation could go on and on, but you can see, Mongrel is hungry and has decided to . . .

# Chapter 13
# Rounding and Nice Numbers

Acer Eraser straightened his tie as Ms. Killion, principal of Mathopolis Elementary School, stepped to the microphone. Acer had spent all morning creating enough chaos to keep those two super meddlers busy for a while. With luck and a little time, he could buy some parent votes by bribing their brats.

An expectant hush filled the huge auditorium as the principal raised her hands for silence. "Boys and girls, I would like to introduce you to Acer Eraser, candidate for Mathopolis mayor. Let's all give him a hearty Mathopolis Elementary welcome."

Polite applause rippled through the auditorium as Acer moved to the microphone. He cleared his throat and gave everyone an almost sincere smile.

"My young Mathopolisians," he crooned. "You know of the recent problems in Mathopolis. You can't buy a new skateboard because your money has no numbers. You do all your math homework only to have it erased. You score for the team, and yet there are no hoorays for you because the scoreboard still says zero to zero. My young friends, I promise you that I will make the world safe for these nice numbers. I make this promise to you because I am your friend." Acer opened his arms in a wide welcoming gesture. "I expect nothing in return except your friendship."

"And votes from your parents," Acer thought to himself.

"But you didn't come to hear me talk. I'm here to listen to you. Ask me your questions. I will answer them openly, honestly, and truthfully."

"Or not," he added to himself.

A tiny curly-haired girl on the front row raised a shy hand. "What's a mayor?"

Acer smiled almost kindly at the little girl. "It's someone who gives candy to a sweet little girl who is brave enough to ask the first question." Acer pulled a huge bag of candy from behind the podium and tossed one to the girl.

Her friends gathered around to get a closer look at her prize, the largest jawbreaker they'd ever seen. Acer's name was on one side of the candy wrapper. "Vote for Acer" was printed on the other.

"Next question," said Acer.

Young hands shot into the air all over the auditorium.

"Do you have a car?" asked a boy.

"No. I drive a scooter." Acer tossed the boy some candy.

"Do you have any pets?"

"Yes." More candy.

"Do you like bananas?"

"Have you ever eaten a worm?"

"Is two plus two equal to four?"

"Can we have more candy?"

"Yes. No. Yes. Yes."

Questions rang out and candy rained down until every child had a pocketful of the sweet treats.

"My gift to you, my little friends," said Acer.

This time the auditorium rocked with thunderous applause.

"Sweet little monsters," thought Acer. "It's a cinch to buy their loyalty. All it cost me was a bag of cavity-causing candy."

"Are there any more questions?" Acer asked, showing them his empty candy bag.

Everyone sagged back into their chairs. There was no more candy, so there were no more questions, except for a fifth-grader in the back row raising his hand.

Acer showed him the empty candy bag.

The boy waved his hand all the harder, like a kindergartener desperately needing to go potty.

"Yes?" asked Acer with a poorly disguised sigh.

"What's your favorite number?" asked the boy.

Acer shrugged helplessly. "How can I choose? Numbers are numbers, and, well, they are all so nice."

"But if you had to choose one," persisted the fifth grader.

"Well," Acer scratched his head as if deep in thought. "I suppose if I had to choose a number it would be . . ."

The auditorium hushed to a complete silence as everyone held their breath.

"Twenty-five."

"Why?" asked the fifth grader.

Acer's smile became just the teeny tiniest bit forced. This kid was really getting on his nerves. How should he know what made a twenty-five special? He'd just blurted out the first number that came to mind.

Quickly recovering, Acer smiled through gritted teeth at the child. "Who am I to say what makes one number better than another? With all these intelligent young minds, I'm sure you can answer that question far better than I." He swept an open hand out toward the children. "What have the brightest of the bright to say?"

"It's divisible by five," shouted one.

"Four twenty-five's make one hundred," chimed in another.

"It's one-fourth of a dollar," said still another.

"It's a really nice number," one more added.

"It makes it easier to estimate calculations," said an older boy.

Acer nodded and smiled and nodded some more. "And there you have it." Then he turned his head and muttered under his breath, "Who would want to calculate anyway?"

What a silly question! Calculations can be used in all sorts of nifty ways to help you estimate. In the last chapter, you learned about truncating, and now we are going to add rounding and nice numbers to your estimating skills.

First, let's compare rounding to truncating because they have a lot in common. They're brother and sister.

Both get rid of digits to make working with them easier. One bites them (truncates) and the other adds donuts (rounding).

Rounding is turning some of the digits into zeros or donuts. The most important digits, the ones that have the greatest affect on the number, are not changed.

Another way to think of keeping the most important digits is to think of each digit as representing a house. Look at the number 234 in terms of houses.

House 200    House 30    House 4

Now, let's say each number represents the rooms in each house. House 200 is a mansion. The next biggest is House 30, with thirty rooms. And of course the smallest is House 4 with only four rooms.

So with the number 234, if you were rounding to the nearest hundred, you would keep the House 200 because that would be keeping the biggest house.

Remember the goal of rounding is to be able to calculate the problem easily and quickly. It is different from truncating because with rounding you have a choice to make.

Remember with truncating you are always making the number smaller because you bite off the extra. With rounding the number can get either smaller or larger.

Let's take an example. Three friends walk into a store to purchase a package of double-dutch jump ropes. The ropes cost $5.79 a package. How much should each have to pay? They don't want to pay more than the other person, and no one has brought a calculator along. So how do you split $5.79 three ways without a calculator?

The problem is $5.79 ÷ 3.

If you truncated you would get 5 ÷ 3, and this is not an easy problem. So how do you round? Simply follow the instructions in the *Rule Book of Rounding*.

Now take a look at how these steps work on the problem $5.79 ÷ 3.

1. **Dizzy Digits.** How many digits can dance in your head before you get dizzy? How difficult a problem can you work quickly? $5.79 ÷ 3 = is too complicated but you can easily work with one digit from each number.

2. **Underline.** Underline what you are going to keep. $\underline{5}.79 ÷ \underline{3} = ?$

3. **Tug of War.** Look closely at the digit to the right of the underlined digit five. How strong is that digit? Remember the tug of war? If both $5.00 and $6.00 pulled the underlined digit, which one would win? Will the digit five get pulled up, or stay the same? The superhero secret to help you make this decision is to look at the number to the right of the

number that is underlined. If it is five or greater, it will pull the underlined number up. If it is four or less, it will keep the underlined number the same. The seven is strong, so the digit increases by one. You add one, and the five becomes a six.

4. **Bring on the Donuts.** Use your donut machine to replace the digits you no longer need with zeros. You removed the seven and the nine, so $5.79 becomes $6.00. You call it rounding up because the number got bigger. If the number got smaller, you would call that rounding down.

5. **Calculate.** $6.00 ÷ 3 = $2.00.

Now all you need to do is to check to make sure everyone has enough money!

Remember these steps: Dizzy Digits, Underline, Tug of War, Bring on the Donuts, and Calculate.

Now, it's time for you to give it a try.

Let's pretend a group of your friends wants to see how many times they can jump. If twenty-four kids each jump 375 times, about how many total times will they have jumped all together? First, ask yourself, "What is the problem?"

$$24 \times 375 = ?$$

Try and solve it for yourself first and then come back and look at the answer.

1. **Dizzy Digits:** You can't do $24 \times 375$, but you can probably do the first digits of each.

2. **Underline:** $24 \times 375$

3. **Tug of War:** Look at the number to the right of the digit you are keeping. For 24, the digit to the right

of the 2 is a 4. It is not strong enough to pull up the 2, so the 2 stays. For 375, the digit to the right of the 3 is a 7, so it is strong enough to pull the 3 to a 4.

4.  **Bring on the donuts:** 24 becomes 20 and 375 becomes 400.

5.  **Calculate:** $20 \times 400$ = about 8,000 jumps

See? That was easy!

Another way to look at rounding is to ask yourself, are you more than half way to the next number, or not? Say you can burp twenty-seven times in a row, and there are two burping clubs in your neighborhood. One is the Timid Twenties Burping Club, and one is the Noisy Thirties Burping Club. You want to go to the club that most closely matches what you can do. Which club would you go to?

The Noisy Thirties Burping Club

Okay, let's review. When rounding, zeros come in and wipe out all the digits except the last one or two on the left. With the last digit or digits you need to decide whether it moves up, or stays the same. Remember that it's a tug of war.

When you are rounding for subtraction, often you only round the number you are subtracting from the larger number.

Take a look.

Say there are 1,345 ants in the mound, and a boy squishes down his foot and wipes out 234. About how many are left?

First, what can you do easily? With subtraction you can just round the second number to make the problem much easier.

You can do it in your head, or quickly on a piece of paper because it's easy when you round the second number.

Now on to nice numbers. Nice numbers are so . . . nice. How are they nice? Do they invite you to their house to play? No, they are nice to work with, which is why you call them nice numbers. When you round and truncate you make nice numbers, but there are other nice numbers besides numbers that end in zeros.

So how do you make nice numbers? You transform numbers into nice numbers in your head.

Let's take an example from addition.

The number of stuffed animals on a girl's bed is the following:

27 stuffed bears
+ 23 stuffed cats
+ 24 stuffed dogs
+ 24 stuffed unicorns

If you were truncating you would do 2 + 2 + 2 + 2 = 8, and put the zero back to get 80.

If you were rounding, you'd do 30 + 20 + 20 + 20 = 90.

To help you work with nice numbers, you need a toolbox, which is actually your mind. It contains all the things you've learned about numbers. So, let's take a look at our nice number toolbox to see if anything can help us with the above problem of 27 + 23 + 24 + 24.

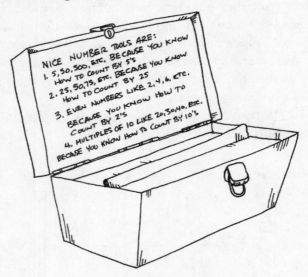

NICE NUMBER TOOLS ARE:
1. 5, 50, 500, ETC. BECAUSE YOU KNOW HOW TO COUNT BY 5'S
2. 25, 50, 75, ETC. BECAUSE YOU KNOW HOW TO COUNT BY 25
3. EVEN NUMBERS LIKE 2, 4, 6, ETC. BECAUSE YOU KNOW HOW TO COUNT BY 2'S
4. MULTIPLES OF 10 LIKE 20, 30, 40, ETC. BECAUSE YOU KNOW HOW TO COUNT BY 10'S

Stop! Back up! Tool #2 can help us figure out how many stuffed animals the girl has. All the numbers in our problem 27 + 23 + 24 + 24 are close to 25. There are four of them, exactly like there are four quarters in a dollar. Twenty-five is a nice number because there are certain things you know about it from money.

$$25¢ + 25¢ + 25¢ + 25¢ = \$1.00$$

So if you make all the numbers 25, 25 + 25 + 25 + 25 = about 100 stuffed animals on her bed.

So the answer to 27 + 23 + 24 + 24 is about 100.
Let's look at another problem.

$$5,501 ÷ 630$$

You could truncate to 55 ÷ 6, or you could round to 5,500 ÷ 600, but neither of these solutions gives you an easy quick answer. So try nice numbers.

Look at the rules in the lid of the nice number toolbox. Which one might help?

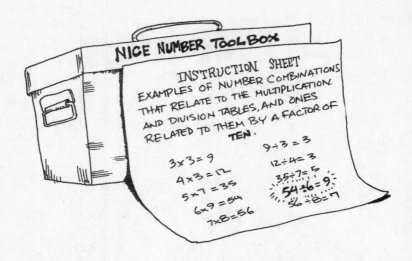

Wait, hold on! 5,501 ÷ 630 = looks something like 54 ÷ 6 = 9 so you can change it to the nice numbers of 5,400 ÷ 600 = 9.

As you learn, you can always add rules to your toolbox. If you learn that 7 + 7 = 14, add that to your toolbox. When you recognize various combinations of ten, add those, too. For example, if Mongrel had two bones for breakfast and eight bones for lunch that would equal ten bones in all. If you have three cats plus seven cats, you have ten, or way too many cats! And four turned-over soft drinks plus six peas shot across the room equals ten messes to clean up!

Okay. You're on a roll, so how about some more?

$$65 + 43$$

What does it remind you of? Think of your toolbox.
It looks a lot like 6 + 4 = 10.
So 65 + 43 is about like 60 + 40 = 100.
Nice numbers also come in handy for subtraction. Again, you can transform only the second number to make

it look more like the first. It's like having a duplication machine.

For example,

$$6,879 - 2,765$$

Leave 6,879 as it is, but look at the second one. The number 765 is not that far away from 879 so transform it. Change 2,765 to 2,879. The problem then becomes $6,879 - 2,879 = 4,000$.

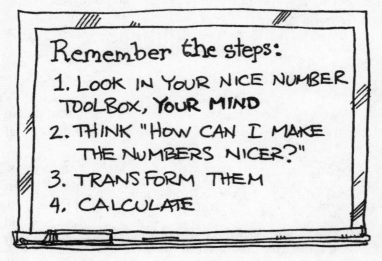

Remember the steps:
1. LOOK IN YOUR NICE NUMBER TOOLBOX, YOUR MIND
2. THINK "HOW CAN I MAKE THE NUMBERS NICER?"
3. TRANSFORM THEM
4. CALCULATE

Use your toolbox to make the following problems nicer, and then solve them.

1. $2,611 \times 398$
2. $612,678 - 598$
3. $2,567 \div 1,289$
4. $221 + 787$

1. Remember quarters $2,500 \times 400 = 1,000,000$.  2. Make the second number match part of the 1st $612,678 - 678 = 612,000$.  3. Make 2.5 into 24 because it works well with 12, $2,400 \div 1,200 = 2$.  4. Think about $2 + 8 = 10$ and change it to $200 + 800 = 1,000$.

And now for some nice number tricks to add to your toolbox. Thirty-three is a nice number to use when multiplying. Don't believe me? Watch! Let's say you invited thirty-three people to your birthday party, and you wanted to give them all fifteen pieces of candy. How much candy would you have to buy?

Okay, here's the trick. When multiplying by thirty-three keep in mind that $33 \times 3$ is about 100 so you can divide the number by three and then multiply by one hundred.

Give it a go.

$$33 \times 15 = ?$$
Divide by 3.
$$15 \div 3 = 5$$
Multiply by 100.
$$5 \times 100 = 500$$

So, $33 \times 15$ is about 500.

How can this possibly work? Because one-third of 100 is about thirty-three.

So if you are ever close to multiplying by thirty-three, you can make your number a nice thirty-three.

Here's one more nice number for the road: fifty. Let's say you have fifty kids in your class, and they each want sixteen pencils.

Here's the trick. When multiplying by fifty or a number close to it, divide the number by two, and then multiply by one hundred.

$$50 \times 16 = ?$$
Divide by 2.
$$16 \div 2 = 8$$
Multiply by 100.
$$8 \times 100 = 800$$

So, $50 \times 16 = 800$.

Wasn't that easy? Now you know three different ways to transform numbers!

All you have to do now is match the following math term with the way to remember it.

Truncating          Donuts and Tug of War
Rounding           Using a Toolbox
Nice Numbers      Biting

Truncating is Biting; Rounding is Donuts and Tug of War; Nice numbers is Using a Toolbox.

# Nice Number Toolbox

- 5, 50, 500 (and its multiples of ten) are all nice numbers because you know how to count by 5's.
- 25, 50, 75, . . . are all nice numbers. You know how to count by 25 because you know quarters.
- Even numbers like 2, 4, 6, and their multiples of 10 like 20, 40, 200, . . . are nice numbers because you know how to count by 2's.
- Number combinations that relate to the multiplication and division tables and the ones related to them by a factor of ten are nice numbers. For example,

$$3 \times 3 = 9 \qquad\qquad 9 \div 3 = 3$$
$$4 \times 3 = 12 \qquad\qquad 12 \div 4 = 3$$
$$5 \times 7 = 35 \qquad\qquad 35 \div 7 = 5$$
$$6 \times 9 = 54 \qquad\qquad 54 \div 6 = 9$$
$$7 \times 8 = 56 \qquad\qquad 56 \div 8 = 7$$

- Combinations that add up to 10 are nice numbers. For example,

$$2 + 8 = 10$$
$$3 + 7 = 10$$
$$4 + 6 = 10$$

- 33 and 50 are nice numbers with multiplication.
- Halving is a useful intermediate strategy in multiplication. For example,

$$14 \times 26 \approx 7 \times 52 = 7 \times 50$$
$$16 \times 56 = 32 \times 28 \approx 30 \times 30$$

Please add your nice numbers to this toolbox!

# Chapter 14
# Getting Closer

Acer laughed with delight. He had Mayor Marbles and Trusty running around in circles.

Today he was disguised as a handyman strolling through the fair grounds munching on popcorn as he observed his latest triumph. Everywhere volunteers and workers hammered and sawed, putting up the last booths for the fair. Some booths leaned, some were on the verge of collapsing, and others were only twelve inches tall.

"Wonderful! Glorious! Marvelous!" Acer thought. Blinded by the need to build everything in time for the fair, the people kept following Acer's Zippo Zappoed instructions. If they had just stopped to think about it, they should have been able to see that what they were building didn't look right.

Chaos had definitely come to Mathopolis, and because of Acer's promise to make the mysterious hooded stranger stop zapping, his campaign was coming along nicely, too.

Of course, as soon as he was elected, the hooded stranger would disappear. But that was the only campaign promise he intended to keep.

"Acer! Acer! He's our man. If he can't do it, nobody can."

Acer smiled at the sound of a familiar jump rope chant and looked around. Off by one of the booths a group of girls were double-dutching.

"Vote for Acer, then you'll see, he'll bring back numbers, One—Two—Three . . ."

In another area, a group of boys dressed as pirates chanted, "Mayor Marbles has to go," they shouted. "Bring on Acer! Yo-ho-ho!"

Acer smiled, finished the last of his popcorn, and tossed the empty bag on the ground. Yes, indeed. The more he erased, the more difficult their lives became and the clearer his mind felt. Without numbers, the people weren't able to measure or calculate, and that was fine with Acer, just fine indeed.

Well, maybe that's okay with Acer, but by now you have a good idea of how truncating, rounding, and nice numbers can help you.

But to be an expert estimator, there are still a few things missing. In the last two chapters you were given a problem and told how to estimate the answer. All of this is good, and it works well with math problems. But how often in the real world do you find things as neat and tidy as they are in a book?

Probably never.

In the real world, no one tells you how to estimate calculations. You have to decide which approach will give you the closest answer.

So which curtain would you tell her to choose? To decide, let's explore the advantages of each way.

But truncating can be bad if you have to throw or give away too many numbers. Say you have two kids, each with gum balls. One has thirty-one gum balls, and the other has sixty-eight. Which kid would throw a fit if you truncated their gum balls?

So which number would be better to truncate, 123 or 189? How about 112,345, or 198,456?

In both cases the first numbers would be better to truncate because they don't have as much to lose.

Perhaps rounding would be the way to go.

Remember, with rounding, you either increase or decrease the value of the number, based on the value of the digit to the right of the underlined digit. With the number 165 you have to choose between decreasing it to 100 or increasing it to 200. Because the number to the right of the underlined one is six it would increase to 200.

Rounding is often the best when some of the numbers would be rounded up.

So how about nice numbers? What are its advantages?

Nice numbers can only be used when the numbers stand out as being nice. There has to be something special about the combination of numbers, so nice numbers can be more difficult to use. (Using the toolbox at the end of Chapter 13 can help.)

Okay, here's a problem.

$$24,987 + 23,857$$

Try truncating, rounding, and using nice numbers to see which way turns out closer to the answer.

Truncate it first. What did you get?

2 + 2 = 4, then put back zeros to make it 40,000. Or maybe you truncated to 24 + 23 = 47, then put back zeros to make it 47,000.

Now round it, and bring on the donut machine.

24,987 rounds up to 25,000. And 23,857 rounds up to 24,000. So 25,000 + 24,000 = 49,000. Or, you could round 24,987 to 20,000 and 23,857 to 20,000 and get 40,000.

Now try to use your nice numbers toolbox.

The exact answer of 24,987 + 23,857 = 48,844. Which estimated answer was the closest?

The rounding answer of 49,000 and the nice number answer of 48,000 were the closest. Why were they closest? Why wasn't the truncated answer closer? Think about what you know about the truncating answers.

With truncation you always lose, and in this case you lost a lot. If you truncate 24,987 to 24,000, you lose 987. If you truncate 23,857 to 23,000, you lose another 857.

Rounding was one of the best solutions because when you rounded 24,987 up to 25,000 you only added 13 and when you rounded 23,857 up to 24,000 you only added 143. You know that's pretty close because you didn't have to round up by very much. But because you rounded up both numbers and you are doing addition you know that 49,000 is a bit high.

Nice numbers also worked well if you chose the nice numbers 24,000 + 24,000. While the first number went down only 987, the second number went up by 143 and helped compensate.

Say this problem had to do with how much money you needed to buy two cars? Which estimate would you use?

Truncating is always low so it is not a good selection. When you rounded, you rounded up, so you know that rounding is too high. When you are checking to make sure if you have enough money, it's always good to have a high estimate.

Because estimates can sometimes be too low and sometimes be too high, instead of giving the answer as about 47,000, you could give a range. You hear people say it all the time. It's between this and that. They are talking about a range. A range is knowing the answer is more than one number, but less than another.

Given the problem 24,987 + 23,857, what range would you give for the answer?

1. You could say the answer is in between 47,000 (truncated answer) and 49,000 (rounded answer).
2. You could say the answer is between 40,000 (truncated answer) and 50,000 (nice number answer).
3. You could write 40,000 < the answer < 49,000. This means in English that the answer is greater than 40,000 and less than 49,000.
4. You could write 47,000 < the answer < 50,000.

WHY DO WE WANT TO CROSS THE ROAD?

Let's try another example with multiplying chickens. Let's say there are 36 chicken houses and 676 chickens in each. They can't decide whether to stay where they are or cross the road.

169

How many chickens might cross the road? When you truncated 36 × 676 what do you get?

36 × 676
36 becomes 3
676 becomes 6
So 3 × 6 = 18 and put back the zeros. When you are multiplying you add the zeros from the first number to the zeros from the second number. 1 zero + 2 zeros = 3 zeros. So 18 becomes 18,000.

Is that answer too high or too low? It's low because multiplying is just adding over and over again. And you took off 3 from the first number and 76 from the second.

Now try rounding. Would the numbers be rounded up or down?

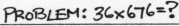

PROBLEM: 36×676=?

SOLUTION: 40 × 700 = 28,000

Ask yourself, is that answer a high estimate or a low estimate?

It's a high estimate because both numbers were rounded up. Also it's high because when you are multiplying you add the extra over and over again.

So you know that the number of chickens that might cross the road is between 18,000 and 28,000. In a math sentence, you write a range as 18,000 < chickens < 28,000. That is a range. (A range of chickens, that is.)

That brings us to a big math word, *compensate*. Compensate simply means to adjust to get a closer answer. We have to look at how changing the numbers up and down affects the answer differently when you add, subtract, multiply, and divide.

To get a closer estimation, sometimes you need to compensate.

To compensate you can either make changes before you estimate, or you can estimate first, and then compensate to make the answer more accurate.

For example, suppose if you were adding forty-six and forty-five and using rounding to estimate the answer. Technically both numbers should be rounded up to become

50 + 50 = 100. But if you want to compensate, you could think it through this way:

$$46 + 45$$

If I add four pennies to 45, I could take away 5 pennies from 45. 50 + 40 = 90.

This is a closer estimation of the actual answer of ninety-one. It's easy to see how to compensate when you add.

Try subtraction. Say Mongrel had ninety-three dog bones and he knows he has eaten forty-five. About how many does he have now?

You might think you could compensate this problem in a similar way to the previous example. You might think, "If I round ninety-three down to ninety, I should round forty-five up to fifty." But that gets you further from the calculated solution because of what it means to subtract. Let's take a look:

When you round down the first number in a subtraction problem, it subtracts more. When the second number is rounded up, it also subtracts more. So with subtracting, if you round the first number down you can compensate by making the second number forty instead of fifty. So 90 – 40 = 50.

You try one. Estimate 196 – 44.

Try to truncate, round, and use nice numbers. Then compensate your rounded answer.

When you multiply and divide, it's much harder to compensate because of their nature. Here are a couple of examples.

$$23 \times 185 = ?$$

Truncate $2 \times 18 = 36$.

Put back two donuts to make 3,600.

You can also compensate by adjusting the second number up since the first number went down.

$$2 \times 20 = 40$$

Put back two donuts to make the answer 4,000. Or you can adjust using $2 \times 21 = 42$.

Put back two donuts to make the answer 4,200.

The calculated answer is 4,355.

Now try this one.

$$185 \div 23 = ?$$

Nice numbers gives you the problem $175 \div 25 = 7$. Because this is a division problem, compensate by choosing a different nice number where each number is increased:

$$200 \div 25 = 8$$

The actual answer is $185 \div 23 = 8.04$.

Be careful when multiply and dividing. These two operations don't react like adding. Because you took off so much from one number you don't necessarily add that same amount to the other. Think, what part or percentage of the first number did you take off or add?

With twenty-three you took three away, a pretty good chunk of pie. So that same size chunk needs to be added to the other number. To compensate you need to increase 185 by the same part of pie and since 185 is in the hundreds it's a bigger number.

There is also a form of compensation called final compensation. It simply means to adjust after the estimation is done. The estimate is updated to take into account how far off the initial estimate is. Think of it as Maverick shining up his tools. You already have a good estimate. You can just make it better.

Take the problem 367 + 542. Truncate it.

3 + 5 = 7. Put back the zeros to get 700.

How would you compensate to get a closer answer?

You can look at how much you truncated off of each number. If you wanted to do this fast what would you do? You could say, "I truncated about fifty off of each number and 50 + 50 = 100. So to compensate I could add another 100 to 700 so 700 + 100 = 800."

Now you try.

1. For the problem 1,263 – 245, think nice numbers and then compensate.

2. For the problem 109 ÷ 23, think rounding and nice numbers.

3. For the problem 51 + 54, think truncating.

4. For the problem 13 × 13, think nice numbers.

1. How about 1,263 – 263 (making the second number look just like the first makes it easy to subtract.) 1,263 – 263 = 1,000. Compensations: If you want to get closer add about how much you increased 245 by to make it 263. That's about 20 so the answer becomes 1,020.

2. You could round it to 100 ÷ 25 = 4. Because both numbers changed by about the same part no compensation is needed.

3. 5 + 5 = 10 and put back 1 zero to get 100. You could compensate by adding back the 5 total that were taken off. Then your answer would be 105.

4. You could use nice numbers of 13 × 10 = 130 or 12 × 12. Compensation: You took away 13 three times if you used 13 × 10, so you could compensate by about 30 or 40. Then you can add that extra to the original estimate of 130 + 30 = 160. If you know 12 × 12 = 144, you would compensate by adding another 10 or a few more.

Fabulous! Things are going better for you than for Acer who has one of the superheroes hot on his trail.

Acer Eraser looked toward the sky and squinted. He'd been watching a yellow speck make tighter and tighter circles.

"Elexus!" he muttered. Disguise or not, it was time to go.

He straddled his scooter and zoomed away. That yellow menace and her silly measuring friend were becoming a real bother.

# Chapter 15

# The "A" Team

*Almost, about,* and *approximately* are words that are used when you are estimating measurements. But why and when do you need to estimate measurements?

What if you were stranded on an island, there was no superhero around to rescue you, and you had a burning desire to know how big the fish was that you caught?

Okay, let's try that again. What if you were stranded on an island without any tools, and you had a burning desire to know the size of the fish that you caught?

Knowing the length or size of something isn't as hard as you might think. And it isn't hard knowing the time or volume of something, either—even without a ruler, cup, scale, or clock.

All you need to do is estimate.

Now, this is a bit different than estimating calculations, as we've been doing, but it is still estimating. Remember, the job of estimating is still not to be precise. Its job is to give you the *wrong* answer.

In case you think estimating measurements is uncharted territory, you are wrong. You already estimate measurements all the time.

See? You are constantly estimating.

Let's look at another example.

Do you remember the boy, way back in Chapter 10, who asked a dangerous question of how much the lunch lady weighed? In the end, he decided it was safer to estimate. His process was the process of estimating a measurement. Whether estimating height, weight, volume, time, or the amount of something, the process is the same.

First, to estimate a measurement, he needed to look at the size—the bigness or smallness of the answer. Would the answer most likely be in ounces, pounds, or tons?

1 OUNCE  7 POUNDS

1 TON

The lunch lady is definitely not the size of a bug or a bison, so our boy settled on pounds.

Next, he started comparing. He searched his mind for all the things he already knows.

A lot of it didn't help, but when he thought long enough, something popped up he could use as a comparison.

So the boy reasoned to himself, "Well, I know the lunch lady doesn't weigh as much as an elephant, and an elephant weighs 14,000 pounds. And I know she's bigger than me, and I weigh 80 pounds. So since the lunch lady weighs more than me, and less than an elephant, I know that she weighs somewhere between 80 and 14,000 pounds. 80 pounds < lunch lady's weight < 14,000."

Now, if this is all the information he had, he would need to consider if she weighed more like he weighed or more like the elephant? Chances are, though, somewhere deep in his mind is something waiting that would help get a more accurate estimation.

He could remember the time he snuck around the corner when his mother was weighing herself, and there in blinking lights on the scale was the number 165.

He might have suppressed the information because it was dangerous. But now it could help because he could compare his mother and the lunch lady. Is the lunch lady as tall as his mom? Are they skinny or fat?

He could say to himself, "Well the lunch lady is skinnier, and a bit shorter than my mom, so I estimate she weighs approximately 150 pounds."

See how easy that was? He didn't even get in trouble along the way.

And that, in a nutshell, is how to estimate a measurement.

Now let's take a closer look at how it's done. Let us go deep in your mind.

When estimating a measurement, something you know is taken and used to figure out something you don't. The place to start with this estimating, as with estimating calculations, is with the bigness and smallness.

If you can get the answer in the right ballpark, you're on your way. Let's try it out on the following objects. Which choices would be associated with the right size?

What would be the general weight of the car?
In the 10's of pounds
In the 100's of pounds
In the 1,000's of pounds

How long would it take a hungry Mongrel to eat a dog biscuit?

> A few seconds
> A few minutes
> A few hours
> A few days

How far would you have to throw a note to reach your best friend's desk?

> Inches
> Feet
> Miles

The car would be in the 1,000 pound area, and Mongrel would only take a few seconds to eat a dog biscuit. And, if you are in the same class and your desks aren't pushed together, you would measure your distance in feet. But if you live in another town, you would measure it in miles.

Now that you have an idea about how to use largeness or smallness, you can estimate by comparing what you know to what you are trying to figure out. Estimating measurements are all about comparing.

There are a couple of different ways to compare. One is to compare directly to standard measurements like feet, cups, or seconds. This only works if you have a good idea of how much a foot is, how much a cup is, and how much a second is.

Let's say you are asked to go outside and find a stick that is about one foot long. You can either take a ruler with you and measure every stick, or you can use the ruler in your head.

The ruler in your head is not a real one. Use a mental image of how long a foot is.

If you don't believe this, try an experiment. Have your friend get a ruler so that you won't look at it. Hold your hands apart so that the distance between your hands is what you believe to be a foot. Now have your friend use the ruler to measure how close you are. Watch Mayor Marbles do it.

MAYOR TRIES TO IMAGINE A
FOOT MEASUREMENT

Never mind—don't watch him. You probably did better than he did. And believe it or not, you have more than a foot in your head. You probably know about how much an inch is, and a yard. If you've ever done any cooking, you know approximately how much a teaspoon is, and you also have a good idea about how long a second is and approximately how much milk goes in a cup. So if your friend asks you for a cup of milk, imagine a cup and use it to measure.

But sometimes this is hard. Most people don't have a good idea of what a pound is, so chances are if someone asks you to estimate the weight of a backpack, it would be hard to come up with an answer. Instead of comparing it to the standard measurement of a pound, you compare it to a weight you already know.

Let's say you are asked to approximate the time. First, think about something you know. Say you just got out of school, and you know school is usually out at 3:00 P.M. So even though you don't have a clock you know it is about 3:00 P.M.

Estimating measurements has to do with how much information you have stored in your mind. If someone asked you how far it was from your house to school, all you would have to do is compare it to a distance you know. If you don't know the distance to any place, or the weight of anyone, it's hard to estimate. So how do you make yourself better at estimating measurements?

Elexus is here to let you in on super estimating tricks.

To increase your mental measuring ability you need to learn more about measuring units. Just as you become a better runner if you exercise your body, you become a better estimator if you exercise your brain. The way to practice is to imagine and then check.

Imagine a cup. Then check. Put a cup of water in a container (without using a measuring cup). Then measure the water and see how close you were. Next do it with a quart and a gallon.

Imagine an inch or a foot or a yard. Draw a line an inch long, a foot long, and a yard long. Now check. Use a yardstick to measure and see how close you were.

Look around and find something in the room that you think weighs about a pound. Now check it by measuring its weight. How close were you?

Okay. Try this. Estimate approximately how long this lightning bolt is in inches.

Now check it with a ruler. How close were you?

Now try to estimate the length of this banner. Remember the lightning bolt. Is it longer or shorter than the lightning bolt? How much longer or shorter is it?

Check it with a ruler. How did you do?

Chances are you did okay because you had the information on how long the lightning bolt was already in your head.

So, to get better at estimating measurements, practice, and keep filling your mind with all sorts of measuring information.

Oops! Mustn't forget Mongrel, who is still stranded on the desert island trying to measure his fish.

Let's go back and help him.

# Chapter 16
# Sampling Estimating

Trusty rushed into Mayor Marbles's office with a stack of papers. "I've just received the latest pre-election popularity poll results from the statistics department."

"Popularity poll? For whom?" asked the mayor as he searched through a drawer, looked under his chair, and shuffled through a pile of papers.

Trusty frowned. He was worried. The stress of dealing with the campaign, the fair, and the mysterious hooded stranger was taking its toll on the mayor. He was even more distracted and forgetful than usual. "It's *your* popularity poll, sir," said Trusty. "To get an idea of how many people plan to vote for you."

"Oh," said the mayor, searching the bookcase and behind the open door. "What does it say?"

"It says," Trusty began.

The mayor brushed past him, walked down the hallway a few feet, and then came back.

"Sir, is something wrong?"

"What? Oh. Yes. I can't find my pen. Have you seen it?"

"Yes, sir. It's behind your ear."

"Oh." The mayor took the pen, and made a note on his calendar. "What were you saying about a report?"

"It's the popularity poll. We randomly asked 1,000 people whether they were planning to vote for you or Acer Eraser for mayor." Trusty placed the stack of papers on the mayor's desk in front of him. "According to our data, your popularity has fallen twenty-three percent in the last two days." Trusty pointed to a row of numbers on the top sheet. "If this trend continues, you stand a forty-three percent chance of losing the election."

"Nonsense," said the mayor. "That report must be wrong."

"But, sir," Trusty dared to argue, "the random sample taken says — "

"I've done my own popularity poll about the election and got vastly different results."

"You what?"

"Take a look at this," said the mayor proudly. He pulled a single sheet of paper out of a stack on the corner of his desk. "My poll indicates that one-hundred percent of the people are behind me." Before Trusty could take a closer look, the mayor shoved the page back into the stack of papers. "So you see, there is absolutely nothing to worry about."

"When did you conduct this survey, sir?" Trusty asked.

"This very morning," the mayor replied with a huge satisfied grin.

"You polled 1,000 people in one morning all by yourself?" asked Trusty.

"I don't have time for that," said the mayor with a wave of his hand. "I asked ten people."

Trusty gave him a suspicious look. "Which ten people?"

"The first ten I saw."

"And who were they?"

The mayor pulled the sheet of paper out from the stack again. "Let's see. There was my wife, my two children, and my barber." The mayor counted off on his fingers. "Then I asked Mrs. Potts next door. You know her. She makes those lovely twinkle tart cookies. The mailman dropped by, so I asked him. And then Amy Nabitz, our City Hall receptionist; the front door guard; and myself." The mayor looked at his fingers, seemingly confused for a moment. "That's only nine. I know I asked ten." He scratched his head and then brightened. "Oh, and I put you down for a yes. You don't mind do you?"

"But, sir—"

"You do mind? You mean you wouldn't vote for me?"

"I'd vote for you, sir, but—"

"Oh, that's wonderful. That's so nice of you. I'd vote for you, too." The mayor suddenly frowned. "But I can't vote for you, can I? You're not running for anything." He looked questioningly at Trusty. "You aren't running for anything, are you?"

Trusty picked up his stack of survey results and sighed. "No, sir. I'm not running for anything. I was just going to say that the people you asked aren't a very good sample."

"They're the best people I know," insisted the mayor.

"Yes, yes, they are great people," Trusty said, "but that doesn't necessarily make them a great sample."

Does it? Or doesn't it? Who is right? Does the mayor have an understanding of sampling, or does Trusty?

Well, a simple jar of gum balls sitting on a librarian's desk will help us find the answer.

Pretend your librarian is having a contest. She is asking each student to figure out how many gum balls are in the jar on her desk. You can't count them all. You can't ask her how many there are. You can't open the jar because the library police will get you. You can pull a number out of a hat, but you won't win the bag of books and goodies. So what do you do?

If you've even been half awake reading this book, you know the answer is to estimate.

½ AWAKE    ½ ASLEEP

So far we've learned how to estimate calculations with truncating, rounding, and nice numbers, and you've learned how to estimate measurements by comparing the size of something you don't know with something you do. Will any of the techniques you've learned so far help you with the gum balls? The answer is no.

So, what will work?

To estimate gum balls, you can use a type of estimating called sampling. It can also help you estimate how many fish are in a crowded fish tank and how many hot dogs you will need to make for a crowd of one hundred people. The only other option is, well . . .

Sampling is the better way, but what is sampling?

Have you ever gone into a grocery store and had someone ask you if you'd like a sample of pizza? You didn't get the whole thing. You got only a small piece.

So the first step in sampling is to find a sample of the total:

1. Count how many gum balls are in a randomly chosen section of the jar.
2. Count how many fish are in a randomly chosen fourth of the aquarium.
3. Ask a few randomly chosen people how many hot dogs they can eat.

If you want to find out about how many dog biscuits are in fifteen boxes, what would you do? First, randomly choose a sample of one box. Next, ask the question: how many dog biscuits are in one box? Then count them.

If you need to figure out how many hot dogs one hundred people can eat, what should you do? The first step is to get a sample of a few people and ask them how many hot dogs they can eat.

Maverick brings up a good point that's important.

# IMPORTANT POINT

When gathering your sample, be sure that your sample is random. You don't want to only ask the biggest eaters or the smallest eaters. Choose who you ask by drawing their name out of a hat.

And it is always helpful to ask more than one person, just in case you get a typical person who is a vegetarian and does not like hot dogs. That would be a very good sample. So try it again with your new knowledge.

So there you have it! You have a good sampling of five randomly chosen people who are coming to your party. But remember if you added up the number of hot dogs those five people will eat, you do not get the total number of hot dogs you need for one hundred people. Those five people are just a sample, five people out of one hundred. How many hot dogs will those five people eat all together?

0 hot dogs + 5 hot dogs + 1 hot dog +
2 hot dogs + 3 hot dogs = 11 hot dogs

After you have your sample amount, the next step is to figure out what part of the whole your sample was. Sampling is taking what you've learned from a small part and projecting it.

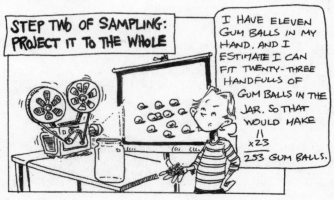

The total number of hot dogs requested in our sample is eleven. Now the next step is to figure out how many samples of five people would be found in a group of one hundred people. Ask yourself how many times five goes into one hundred.

Write it as $100 ÷ 5 = 20$. So to get the total number of hot dogs, take the number of hot dogs the sample group would eat, eleven. Next multiply that by the number of groups it would take to get one hundred people if each group had five people. The number of groups would be twenty.

a.  $100 ÷ 5 = 20$
b.  $11 × 20 = 220$

You need about 220 hot dogs for one hundred people.

Next take a look at the fish tank. First, get a sample and ask a question.

For your sample, take one-fourth of the tank and ask the question: How many fish are in one section?

So randomly choose a section of the tank and count the fish in it. What did you get?

Upper left corner eleven fish. Upper right hand corner twelve fish. Lower left hand corner thirteen fish. Lower right hand corner eleven fish.

The number you counted will depend on which fourth you selected, but in case you think this is a problem, remember this is an estimate.

Now take what you learned from this small part, and project it onto the whole.

There are four sections, so your sample was one-fourth of the total amount. So to find the total amount, multiply the amount in the one sample or section by the number of total sections. So eleven fish in the sample times four sections equals forty-four estimated fish.

$$11 \times 4 = 44$$

Now you try a bit of sampling.

1. If you have a package of 200 seeds, how would you figure out about how many seeds would sprout out of those 200 seeds?
2. How many red M & M's are there in a one-pound bag of M & M's?
3. How would you figure out how many people Elexus helps each year?

1. Plant ten seeds and water. Count the number of seeds that sprout. Multiply that number by twenty to get the total number of seeds that would sprout out of 200.
2. Take a handful of M & M's. Count the number of M & M's in that sample. Figure out about how many handfuls are in the entire bag. Multiply the number of handfuls in the entire bag. Multiply the number of red M & M's in the handful by the number of handfuls in the entire bag.
3. Count the number of people that Elexus helps in an average week and multiply it by fifty-two.

Finally, we come back to the mayor and Trusty and what makes a good sample.

"What do you mean the people I interviewed aren't a good sample," said the mayor. "They're the best people I know."

"They were all people who would vote for you," Trusty said trying to explain.

"That's what makes them a great sample," said the mayor.

"They weren't randomly chosen," Trusty said, hoping he wasn't hurting the mayor's feelings. "That's actually what makes them a bad sample."

The mayor frowned, thought, and then his face broke into a broad grin. "Bad for Acer, maybe, but not bad for me."

Will Acer or Mayor Marbles triumph in the election? Read on to find out what happens!

# Chapter 17
## Friends or Enemies

"We've got to stop this," Elexus said, staring at a huge new "Acer for Mayor" billboard erected just outside the fair grounds.

Mongrel moved in closer and sympathetically rubbed against her leg. She reached down absently and scratched behind his ears.

"We need to set a trap," said Maverick, sitting at the picnic table beside her.

"What kind of a trap?" Elexus asked.

"I have an idea," said Maverick. "But we will have to work together."

He looked around, motioned for Mongrel to join them, and waited for the dog to jump up on the bench. Maverick lowered his voice, leaned in, and quietly outlined his plan.

"That just might work," she said.

Mongrel barked his approval, too.

"He'll never see it coming," Maverick said, smiling.

"And the people will want to keep Mayor Marbles," Elexus added.

"Then it's off to the rescue," Maverick said, fingering his tool belt. He held up his hand to Elexus, and she gave him a high five.

"Measure it or estimate, both are great to calculate. Use them wisely. Don't compete. As a team, they can't be beat," Elexus chanted.

"Not bad," said Maverick. "Not bad at all."

And they both laughed.

In this book you've learned about both measuring and estimating. You've learned about different types of estimating like sampling, estimating calculations, and estimating measurements. You've also learned how to measure time, distance, weight, volume, and temperature. Now the questions are:

1. When do you get out your calculator, or count, or measure, and when do you estimate?
2. When do measuring and estimating help each other out? In other words, are they friends or enemies?

Let's start with when to measure and not estimate. Please don't just estimate when a precise answer is needed. Races are not estimated. They are measured.

The store doesn't estimate how much you owe when you go to pay. They calculate it because they need an exact amount.

You can't estimate the answer on a math test. Again you calculate it. You should always use estimation to check your answer, though, and sometimes an estimate is all you need for a multiple-choice test.

Use estimation to check your answers, but don't estimate the answer. You wouldn't feel good about a doctor who estimated how much medicine you should take, right?

Carpenters always have a measuring tape handy because things don't look the same when they aren't measured.

And sometimes recipes need to be measured when they call for exact amounts.

But sometimes estimating is best. One of the biggest advantages of estimating is speed. In most cases, you can estimate a calculation before you can do the calculation (unless you have a calculator in your back pocket.)

There are also other times when you have to estimate. You must estimate when it's a future time event. You can't measure the future.

Madame Sylvie may be able to see the future, but you can't. You don't know the exact time a storm will hit. You don't know the exact time the bus will arrive home or the exact time you'll finish your homework.

Also, estimate when measuring would be dangerous. You might want to estimate the number of teeth a lion has or the height of your roof. It also may be dangerous to try to count the gum balls.

When something is happening too fast to count, estimating is the best choice. Speed, as you learned earlier, is a major advantage of estimating. You can estimate the number of people that run into the fair because they move too fast for you to count.

On your math test when you want to check an answer that has already been calculated, estimate.

That's the way to use estimating on a test! Or you can use estimation to narrow down the answers on a multiple-choice test. You simply estimate and eliminate the impossible answers.

Take the question $98.5 + 287.32 + 601.23$.

a. 1,532.05

b. 987.05

c. 545.05

d. 1,376.05

Notice that all the answers are very different. Do you need to take the time to calculate? You could, but it would be faster to estimate.

Truncate or round or use nice numbers and choose your answer. What answer did you get?

Truncation: $9 + 28 + 60 = 97$ and add back a zero to get 970
Rounded to the nearest hundred: $100 + 300 + 600 = 1,000$
Nice numbers: $100 + 300 + 600 = 1,000$
All these estimated answers get you close enough.
Your answer would be b) 987.05.

When you don't have the tools to measure, estimate. Everything you need to estimate is with you. All you need is a brain, and fortunately you carry it with you at all times.

When there is not enough information, estimate. With sampling there is not enough information or time to ask everyone how many hot dogs they can eat.

But often we estimate for only one reason: there is no need to have an exact answer. There is no need to know exactly how much popcorn is in the bag. There is no need to know exactly how many ping-pong balls will fit in your room. There is no need to know exactly how long it took you to walk a mile unless you are in a race. Remember that, when this is the case, special words are used.

Now for the second question: Can measuring and estimating help each other? The answer is a very definite yes!

People make mistakes all the time when they are measuring and calculating. Estimating can help us correct those mistakes.

Sometimes our calculations on a test are wrong, and sometimes we make an error when the calculation is important. Estimating can help us determine if we need to do the calculation over again. Say you were calculating: 24 + 39 + 78.

$$
\begin{array}{r}
24 \\
+ \quad 39 \\
+ \quad 78 \\
\hline
= \quad 121
\end{array}
$$

Is the answer correct? Use estimating to figure out if you need to recalculate the answer. What would you do to estimate this problem? You could round. 20 + 40 + 80 = 140. The answer doesn't look quite right. You may need to do it again.

$$
\begin{array}{r}
24 \\
+ \quad 39 \\
+ \quad 78 \\
\hline
= \quad 141
\end{array}
$$

Ah, ha! You were off!

Also, if you measure often, you can improve your ability to estimate. Since estimating is based on what you know, by improving your knowledge of how tall, how heavy, how long things are, it improves your ability to estimate measurements.

Now let's see if you could do Mongrel's job and call the appropriate superhero for each problem you face. Call either Elexus Estimator to estimate or Maverick Measurer to measure for each of the problems below.

1. You are going to your friend's house to watch a movie. Your mother asks you how long you'll be gone. Who ya gonna call?

2. You are following a recipe and it calls for two cups of flour, a half cup of applesauce, and one teaspoon of salt. Who ya gonna call?

3. Your math assignment for today asks you to figure out the approximate height of the tallest building in your town. Who ya gonna call?

4. Roughly how many worms do you have if you have twenty-seven containers with twelve worms in each? Who ya gonna call?

5. You are building a model airplane for a science project. The directions call for a wing that is eight inches by three inches on a piece of wood that is one half inch thick. Your mother will cut it out for you, but you need to draw it on the piece of wood first. Who ya gonna call?

6. Your dad calls from the grocery store and asks you how much milk is left. You go to the refrigerator. Who ya gonna call?

7. Your dog is sick and needs medicine. The veterinarian needs to know how much your dog weighs because the amount of medicine you give will be determined by the weight of the dog. Who ya gonna call?

8. The doors open to let you on the elevator. You see a big sign saying it will not carry over 1,000 pounds. On the elevator already are four tall men, and three average-sized women. Should you get on or wait for the next one? Who ya gonna call?

9. You want to mail a letter to your best friend who moved away. It's a four-page letter, and you don't know how much postage you need. You want to make sure it gets there, but you don't want to waste any money. Who ya gonna call?

10. What is $23 \times 3$? Who ya gonna call?

1. Elexus; 2. Maverick; 3. Elexus; 4. Elexus; 5. Maverick; 6. Elexus; 7. Maverick; 8. Elexus; 9. Maverick; 10. Maverick.

# Chapter 18
# High Fives

The mayor practically danced into his office. "It's the most successful advertising promotion City Hall has ever initiated. It's all anyone talks about, adults and children. Everyone plans to participate. It's absolutely the best idea ever, and everyone absolutely loves it."

Trusty beamed at the mayor. "Do you realize you just used "absolutely" twice in the same sentence?" he teased.

"Absolutely," said the mayor.

"Sir, whenever you're ready," said Trusty, "we do have a few last minute details to attend to."

The mayor nodded. Although he tried his best to look serious, he eyes still twinkled with delight.

Trusty handed the mayor a list. "We'll need to add on to the tent to handle the overflow, the pulleys need to be oiled and tested, Maverick has requested an additional 300 feet of rope, and Elexus needs an estimate of the expected attendance so she can make sure everyone will have a clear view."

Later that day, Acer Eraser shifted the bundle in his arms as he waited in the long line at the Mathopolis Bizarre Bazaar ticket gate. There were even more people at the fair today than yesterday. Despite all of Acer's attempts to ruin the fair, someone had come up with a last-minute scheme to save the day. "But I'm not beat yet," Acer muttered to himself. He pulled his hat further down over his silver hair and settled his sunglasses on his nose. It wouldn't do to be recognized. Not when his goal was nearly within his reach.

Arriving at the gate, he handed the ticket taker his money. "One ticket, please."

The girl counted his money. "You're paying full price?" she asked, giving him a confused look.

"Yes," grumbled Acer.

"But you can get five dollars off the price if you bring something that has a five on it."

"I know," muttered Acer.

"It doesn't have to be anything big. A key chain with a five on it, or a business card, or a model car." She looked sadly at Acer as if he were too dull to understand. "Anything with a five. Anything at all."

"I don't want to bring something with a five on it," Acer growled.

"But Mathopolis is trying to break a record. We're assembling the largest collection of fives in the world. Don't you want to help? Don't you want Mathopolis to be in the Numbers Book of World Records?"

"I do *not* want to help. I do *not* want Mathopolis in the Numbers Book of World Records," Acer snapped. "I do *not* want a discount. Now give me my ticket."

The young girl looked like she was about to cry as she handed him his full-priced ticket.

Acer slipped into the fair grounds, and shifted the bundle to his other hand. Disgusting! All these happy people! All the excitement! And all because of what? A silly old number five.

Everyone seemed to be carrying a five of some kind. Big fives. Little fives. Medium-sized fives. A feathered five. A sea shell five. An ice-sculptured five. (That one left a drippy trail.) There were also fives on shoe laces and fives on sweatshirts. A couple of kids kept giving each other high fives and laughing as if each hand slap added to the city's fives collection. It was enough to make Acer dizzy.

All the five carriers headed in the same direction, toward the huge tent erected in the middle of the fair grounds. Well, he would soon put a stop to this nonsense for good.

When he got close to the tent, Acer ducked behind the Shoot Some Hoops booth.

And then he waited.

After a few minutes and an ear-splitting series of squeals and squawks, the loudspeaker system came to life. "Attention." Squeal. Squawk. "May I have your attention, please?" Squawk. Squeal. Buzz. "The opening of the greatest five exhibit in the world is about to take place. Please make your way to the bleachers for the unveiling ceremony."

The excited chatter of the people rose to a higher pitch as they hurried past Acer's hiding place.

Perfect, thought Acer. The bleachers containing all those horrible five lovers are on the other side of the tent. While they make long, boring speeches congratulating each other, he'd be able to sneak into the tent before the ceremony ended. Once inside, he could indulge in a Zippo Zappo Eraser marathon. Those disgusting fives would be history. Oh, how sad for the people. Oh, how humiliating for Mayor Lostis Marbles and Crusty Musty What's-his-name. Acer rubbed his hands together in anticipation.

"And oh, how delightful for me!" he thought.

The noise of the passing people soon receded into the distance. The only living thing in sight was a shaggy mongrel dog gnawing on a bone near the entrance of the huge tent. Unrolling the bundle he carried, Acer exchanged his hat and sunglasses for the hooded disguise he'd been using to terrorize the city. With a flourish and swirl of his cloak, he stepped out from behind the booth and crept toward the enormous tent.

Nothing moved. Even the breeze failed to stir the dust. He heard laughter and applause and more squeals and squawks coming from a distance. Someone must be making a speech.

Arriving at the tent, he looked all around. No one was in sight. Even the mongrel dog had disappeared. He lifted the flap and peered inside. A triangular slash of light illuminated the floor. Just beyond, he could make out dozens of tables and shelves littered with five-fanatic junk. A few fives even hung from the tent poles. Dim light filtered through the tent and cast eerie five-shaped shadows on the floor.

Acer let the flap drop behind him and edged his way past a five-shaped plant stand and over to the tables. This was as good a place as any to start. Pulling out his Zippo

Zappo Eraser gun, Acer grinned in evil anticipation. It was a Zippo Zappo dream come true.

"Ready!" he said to himself. "Aim!" Acer raised his weapon.

A dog howled.

"Fire one!" said a guy's voice from the darkness.

Something snakelike closed around Acer, pinning his arms to his sides. He tried to run.

"Fire two, and three!" said a girl's voice.

More snake things closed around Acer's ankles and legs. He stumbled. He tripped. He crashed to the ground. All his plans crashed with him as he lost his grip on the Zippo Zappo Eraser gun and it bounced away. His hood fell across his face. Blinded but refusing to give up, Acer slithered his way along the ground. "I have to get out of here," he thought. "I have to hide."

"Ready the pulleys," said the guy.

"On the mayor's signal," said the girl.

"I know those voices," thought Acer.

Squeal. Squawk. "And now," announced the mayor over the loud speaker system, "It's time for the unveiling!"

"Pull!" said the girl.

The sides of the tents whooshed open, flooding the entire enclosure with sunlight.

"Behold, people of Mathopolis," said the mayor, "Your record-breaking collection."

Applause, cheers, and whistles rippled all around Acer.

"And a little something extra for your enlightenment," the mayor continued.

Someone pulled back Acer's hood. He blinked at the sudden brightness.

People crowded into the bleachers by the tent gasped in surprise!

The owners of the two voices he'd almost recognized held the ends of the ropes tying him. It was those two meddlesome kids in those ridiculous costumes.

Then the people yelled almost as one. "It's Acer Eraser!"

"How could you?" they shouted.

"We trusted you!"

"You betrayed us!"

"You lied!"

The mayor stepped forward and signaled for silence. "Acer Eraser, I hereby arrest you in the name of the great city of Mathopolis."

"On what charge?" Acer demanded, unwilling to believe he'd lost.

"Conspiracy against the people, plotting to overthrow the government of Mathopolis," said Mayor Lostis Marbles. "Aggravated assault on helpless numbers," he added.

"And . . ." The mayor held up a crumpled popcorn bag. "Littering." The mayor waved toward two burly police officers. "Take him away. Lock him up and throw away the key."

The people cheered as Maverick and Elexus handed Acer over to the police.

Mayor Marbles hushed them with outstretched arms. "I declare this fabulous fives exhibit an official world record!"

The people cheered and streamed into the tent to admire and ooh and aah.

"What do you think, Mayor?" Maverick asked. "My trap worked."

"And not a five was touched," added Elexus.

"It was a great plan," Mayor Marbles said as he shook both their hands. "I can't thank you enough."

"No problem," Maverick said. "Elexus estimated the size of the tent." He gave her a proud grin.

"But Maverick measured the ropes so he'd know exactly how far to throw," Elexus added, returning his smile.

"A true team effort," said Trusty Dusty. "Good job!"

A group of children zoomed past chanting. "Marbles! Marbles! He's our man. He's a great mayor, and I'm his fan."

Not to be outdone, a group of boys shouted, "Our Mayor Marbles runs the show. He stopped Acer! Way to go!"

Mayor Lostis Marbles wiped a tear from his eye.

Trusty Dusty scurried about the fair organizing and making lists.

Their mission accomplished, Elexus and Maverick poofed back to their superhero homes.

And Mongrel curled up beside a hamburger stand, chewed on a meaty bone, and continued watching the world from behind his mop of shaggy fur.